A
Year
Full
of
Flowers

*To Josie Lewis and
Arthur Parkinson*

A Year Full of Flowers

Gardening
for all seasons

Sarah Raven

BLOOMSBURY PUBLISHING
LONDON · OXFORD · NEW YORK · NEW DELHI · SYDNEY

Photographs by
Jonathan Buckley

Contents

Introduction

My love of the natural world came from botanising trips I went on with my dad as a child, but when I started gardening as an adult, my interests took a rather different turn.

He could spend hours carefully walking through a bog or meadow in search of a tiny clump of rare orchids or clubmoss that had been recorded there. That bored me, but I loved the abundance of flowers we often found as incidentals in these spots – anemones sweeping across woodland floors in March, fragrant bluebells in April, carpets of snake's head fritillary in the Thames Valley in May, and viper's bugloss and clouds of valerian on the shingle coast of Kent in June. If you put your foot down anywhere in these places, you'd be crushing a carpet of colour. That's the thing I found instantly inspiring and happy-making – flower and colour – and I still do. It's that lift, that same childlike pleasure I'm looking for from my garden.

I remember a designer friend of mine saying gardens had to work in black and white, that they were only 'good' if they had strong bones and architectural form all through the year. Colour was froth, a temporary distraction, and should be the least of one's priorities. I felt cowed at the time, aware that this was the opposite of the garden I was making.

More than twenty-five years on, I'd say that a strong structural design is only part of it. Yes, you need the bones, but it's the fleshing out with colour that gives me joy. I love topiary and hardworking evergreens and interesting foliage, the sort of thing traditional structuralists think 'make' a garden. I've worked on the bones of our garden, Perch Hill, over the years.

We have some nice buildings which we've gradually restored, gardening the spaces between them. And we do have a small lawn, a little rice with our curry; I value its calm, but it doesn't excite me.

What I love is different. I like plants jam-packed, as you might get in nature, and I'm sure that's why I love plant trials, not just for the experimentation and for the lessons learnt, but for the visual results – one plant, in many different forms, repeated over a large area as if it has spread there naturally.

Nature often produces a single flush of flower, but in a garden, one needs succession. That's what I've been working on at Perch Hill. We have different plants taking centre stage from one week and one month and one year to the next. I use so many annuals, biennials, tender perennials and bulbs, that the 'froth' changes all the time. In fact, in parts of the garden, if you stood in one place in June, you'd hardly recognise it the following June as so many of the plants will have changed. The bones are the same, but the colour-givers are often new. It's these waves of colour, rolling up into the garden one after another that keep me dreaming and thinking of new ideas.

Abundance is pretty easy to achieve during the full-on growing season when you have families like tulips, roses and dahlias up your sleeve, but there are almost equally strong performers in February and November, just not so many of them. That makes them doubly worth planting. I've spent years trying to discover as many of those early and late performers as possible, to help maintain our high colour bar.

I want every direction you look in at Perch Hill to be like a May ball in full, dressed-up parade. Even in the trials, I want a sociable mix of different plants, not just one or two. I love playing with combinations until I find one that works on every level: structure, floweriness, staying power and a spirit-lifting surge of colour. Those are the combinations you'll find here. There are no great rarities, no-tricky-to-grow challenging plants, but easy, life-enhancing colour and flower performers:

Perch Hill

We moved to Perch Hill in 1994 from London and found a rather
ramshackle ex-dairy farm with a lot of concrete, corrugated iron
and a small garden with a goldfish pond on the south side of the
house. Since then, we converted the farm into an organic 90 acres,
putting in new hedges on old lines, trying to encourage wildflowers
into the meadows and introducing our own herd of Sussex cattle
and a flock of Romney-cross sheep.

Like most farmhouse gardens, we have different areas given
over to different purposes. The first garden I made was the cutting
garden, and I used it mainly to trial plants for picking. I have now
divided this area into two, one filled with perennials, the other
with annuals, hardy and half-hardy, and biennials, with two or
even three different crops in the same square metre of soil in one
calendar year.

Dahlias and chrysanthemums are important to me and we made
a trial garden just for them. Every year new varieties go in, gathered
from my trips to Holland and elsewhere the previous autumn.

My younger daughter Molly
and I having a picnic in one of
the Perch Hill fields.

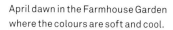
April dawn in the Farmhouse Garden where the colours are soft and cool.

August dawn in the dahlia trial beds in the Perennial Cutting Garden, where pretty much any colour (bar white and very pale) is included.

The western-most garden is also a trial garden, where we test out new annuals from seed, and recently I put in an area dedicated to experimenting with roses that are good for picking.

On the slope, we grow edible crops in an ornamental way, with lots of edible flowers, as well as salads and herbs in patterns of contrasting shapes and colours for the Perch Hill school kitchen.

There are also four purely ornamental gardens that are here for themselves and not for harvesting. There's the Oast Garden, which is an extravagant mix of colour and structure (with salvias, cardoons, artichokes, dahlias and gladioli). Oasts traditionally have wooden balconies extending almost the full length of the building. These are called 'greenstages' and are where hops would be stashed before being loaded into the kiln. We restored ours and cover it with pots all year.

We also have the Farmhouse Garden that's designed to have a calmer feel, with perennials and roses in soft pinks, mauves and blues. On the south side of the barn, there is the Rose and Herb Garden. The herbs are the best culinary varieties, while in the outer beds we have roses chosen for scent and beauty. In 2016, we created our Dutch Yard on the north face of the farmhouse; it's a sort of hybrid between a Dutch painting and a traditional front garden you find in the Netherlands. We bricked the

The Annual Cutting Garden crammed with jewel-coloured tulips.

whole area so we could fill it with pots and classic Dutch-yard plants such as amelanchier, hydrangeas, a mulberry and a catalpa tree.

At the most southern point of the garden, beyond the cutting garden, we have a chicken run and a damson orchard, with a short avenue of willows, all sitting in a wildflower meadow which is filled with colour from spring to autumn.

Plant trials

Gardens such as Great Dixter, the annual border at Nymans and the cottage garden at Sissinghurst have been an inspiration in their full-on and confident use of strong plant colour, and trips to trial fields to see tulips, roses, annuals and dahlias are hugely important to me. Visits to these places have me brimming with ideas and things I want to try in my borders and trial gardens.

I love a trial and have been experimenting from the moment I started gardening. I was drawn to the plant trials run by the RHS, which had many varieties of one family planted cheek by jowl, so you could really see the difference between them in one moment, but also, importantly, over weeks and months, or with trees, shrubs and perennials, over several years.

In many ways, Perch Hill is a trial garden in its entirety. It's about trying things out, seeing what works, what fails, what thrives. I change things around all the time, much more frequently than most gardeners either want to or have time for.

Four of us work in the two-acre garden and about half of our time is spent propagating, planting and assessing what we have on trial. Trials appeal to the doctor-scientist in me, as well as the creative. The selection is always subjective and personal, and largely to do with the shapes and colours that appeal to me, but we analyse objectively. We look at which cultivars or hybrids perform for the longest, flower most prolifically, are least affected by pests and diseases, and have added characteristics such as perfume. As much as we can, we monitor, measure and record everything, so at the end of the trial, we know which we love, and we also have concrete performance information underlying our preferences. I've included lots of our trial results here, so you can see how we've made our selections.

I have grown or trialled thousands of different plants over the decades and I am convinced these selections are the very best. They are the plants and combinations that make me want to sing.

Looking from the Dutch Yard into the Oast Garden, which is presided over by a 300-year-old oak. Moving from April into May, there's more and more colour, as all of our pots are full of tulip bulb lasagnes.

January & February

Gardening in January and February can be pretty unappealing. It's often raining, and when you live on heavy clay as I do, every time you come back into the house you bring half the garden with you.

Thank goodness there are a few plants that can brighten our gardening life and add some colour to these cold, dreich days. I'm talking small and delicate here, not chunky and robust. I think to myself: look down, not out, pull things close, collect them in key spots, don't scatter, just appreciate perfection in one flower rather than armfuls. There's something rather lovely about focusing closely at this time of year.

I have small jugs and tiny vases for this period and a few years ago my husband Adam gave me some old, beaten-up pewter platters. I love filling them with miniature vases and bottles in richly coloured glass and adding a stem or two to each one. This makes a flower arrangement, not in the same vase, but in the mix of things.

At this time of year, when it's cold, cut flowers don't need to go straight into a bucket of water. I simply walk round the garden with a basket or trug and pick sprigs of whatever I can find. In January, it may be just hellebores, but possibly snowdrops, Algerian iris and hazel catkins.

There are some new varieties of hellebore that flower from the start of winter, with the slate-crimson 'Maestro' a favourite, joined by its brother, 'Merlin', at the beginning of February. These particular hellebores are not only glamorous in the garden and in containers, but if you cut more mature stems (where one flower is already at seedpod stage), and sear the stem ends for ten seconds in boiling water, they make excellent cut flowers. After searing, lay them flat in a sink of cold water and leave overnight, and don't cut the seared stem off when you arrange them. (See page 229 for more on conditioning cut flowers.)

Cyclamen coum, *Anemone coronaria*, winter aconites, crocuses, *Fritillaria raddeana*, violas and polyanthus are also appearing by early February. I pick single stems or mini bunches, almost always sticking to one type of flower in each bunch. Once I have a small handful, I tie them together with a rubber band and put them in a trug. That way, each bunch stays together rather than falling into a chaotic shemozzle. Once inside, I plonk them in whatever vase I'm using, and only when the delicate stems are being held by the neck of the vase do I cut the rubber band. It saves a lot of fiddle.

As well as cut flowers, I love to arrange a collection of small pots on a metal table or in our tiered plant stand by the front door. Saxifrage is great for this: you'll see it in window boxes everywhere as you walk through towns or cities during winter – it can be ubiquitous at this time of year, so it's easy to feel snooty about it, but don't.

Below One of our plant theatres packed with forced bulbs, winter-flowering saxifrage, primulas and violas.
Opposite *Helleborus × ericsmithii* 'Merlin'. This is brilliant for winter pots in dappled shade and is excellent for picking.

We grow *Iris reticulata*, winter-flowering pansies and evergreen echeverias (which are not hardy, so need to be brought in if a frost is forecast), as well as *Primula malacoides* and *P. forbesii* – their January powder-puffs of pink flowers are invaluable.

Then there are the early hybrid primroses and polyanthus that we place by the back door. 'Stella Champagne' is my favourite polyanthus, which we have in a huge panel in the Annual Cutting Garden, as well as in a series of containers; in winter I can visit it with my trug week after week. It's ideal for the soft, cashmere-jersey colours of apricot and peach, but for richness, choose 'Stella Regal Red' and 'Stella Neon Violet'. These have flowers in a cluster together at the top of a decent length of stem and so are much more noticeable in a border, and they make small scale but glamorous vases of cut flowers.

I also like the doubles, such as 'Strong Beer' and 'Cobalt Blue', compact, dome-formers, a bit too neat – they get lost in a flower bed, but are ideal for pots, massed several plants together in a velvet carpet. If sown in early summer, they flower on and off throughout winter. If we have a few very cold or wet days, they falter, but perk up again with a bit of sun.

Opposite The Annual Cutting Garden with a good block of polyanthus 'Stella Champagne'. Below A cup from Mexico full of February flowers – *Iris reticulata* 'Fabiola', pulmonaria, polyanthus 'Stella Neon Violet' and 'Cobalt Blue', crocus and the first *Anemone blanda* and *A. coronaria*.

My birthday is at the beginning of February and I love filling the house with friends and lots and lots of flowers by everyone's beds and all over our dining table. That's a challenge at this time of year, but I've learnt about the plants that flower pretty reliably and are good in a vase, and one of them is polyanthus – I turn to it for abundance more than anything else come this time of year.

Spring bulbs, forced so they flower a few weeks early, are another good option. My favourites are *Scilla mischtschenkoana* and *Crocus tommasinianus*, as well as *C. minimus* 'Spring Beauty' – all three can be encouraged to flower by February if planted in September. For large pots, we have a clutch of sweet-smelling beetroot-coloured hyacinths including 'Woodstock' and 'Purple Sensation'. It's the blue and soft pink hyacinths that seem to have the strongest fragrance, with 'Anastasia' forced a month early another February boon.

I move all of these plants inside as they come into bud, to give us something cheery on our kitchen table. However, polyanthus in particular prefer damp and cool conditions and can get leggy and less floriferous if left in the kitchen's warmth for a few

Below Our aromatic rosemary bench looks and smells good.
Opposite An ink bottle arrangement on a pewter dish with *Viola* 'Aquarelle Flambé Toscana', *Crocus* 'Snow Bunting', *Helleborus* × *ericsmithii* 'Maestro' and primula.

days. But if you allow them just a brief visit, they are perfect for bringing winter colour and scent indoors.

I like to combine all these miniature plants with aromatic and perfumed shrubs that flower at this time of year. I grow six different types of rosemary, four of which start to flower in February. Then there's the sarcococcas and daphnes, which exude their perfume in wafting clouds. I pick sprigs from these to bring fragrance inside.

The year starts quite delicately, but can still be full of colour, scent and intensity – concentration is the key.

Opposite Hyacinth 'Purple Sensation' and winter-flowering saxifrage (*S.* × *arendsii* 'Alpino Early Picotee').
Below A Moroccan bowl full of polyanthus (the best dusky colours) topped with dried oak leaves.

Winter-flowering irises

I love the shape of an iris – like a beautiful plumed tricorne hat in the finest silk-velvet – as well as the dabs and strokes of colour on its falls. They're splendid at any time of year, but particularly precious and wonderful when they flower in winter.

I was brought up with the Algerian iris (*Iris unguicularis*), on my parents' doorstep in Cambridgeshire. Their leaves always looked a mess and I wondered why these scruffy plants had made it to such a prominent position. But then, with everything else in hibernation, out they came, two or three flowers appearing and unfurling each week for months – strange cigar-like, pale mauve tubes. My mother used to pick them and put them in a small sherry glass on the kitchen table and I loved them.

My next iris memory is of a family holiday to Corfu when I was ten. My dad had emphysema and often got bronchitis, so trips to warm places were ideal. He and I would go off into the hills on botanising expeditions. We often drove rather than walked (because of his lungs), and he taught me to botanise pretty effectively at 20 miles an hour.

There was a carpet of *Anemone coronaria* along almost every road – I'd never seen wildflowers like it. My father had this one-in-a-hundred picking rule: if something was super-abundant, it was alright to pick a few. There were so many anemones, in a wonderful range of colours from magenta to white, that we stopped so I could pick a bunch. It was then that I spotted my first ever widow iris, the glamorous and exotic, small jet-black snake's head (*Iris tuberosa*), with its perfectly contrasting olive-green heart. What a moment: swallowtail butterflies landing around me and this beautiful iris radiating from a rocky crevice.

Closer to home, early bulbous *Iris reticulata* is fantastic as it flowers in February and is very easy to grow. I try to plant as many as possible in autumn so there's something to enjoy at this dull moment of the year. Both Algerian iris and snake's head iris take a few years to settle in, but not the reticulatas. Even if you plant them as late as November, flowers are guaranteed in February.

Iris 'George' is the top layer of these bulb lasagnes, flowering in February (see p365).

I think my favourite is the richest, tallest and darkest of them all, a highly scented variety called 'Purple Hill', with the purple 'George' and the soft blue 'Harmony' coming in joint second. Then there's the gold iris, appropriately called 'Happiness' and similar to *Iris danfordiae*, which also flowers in February. The prominently veined 'Katharine Hodgkin' is slightly cadaverous, but I like its strangeness. Most irises have the added bonus of a delicate perfume. Last year, we tried a new one, hideously named 'Scent Sational', that smelled strongly of those old-fashioned sweets, Parma Violets.

I love growing reticulatas in shallow bulb trays. I forage a few bits of carpet moss from the woods to fill the gaps between the emergent stems as they grow. Over the top, I scatter dried oak or field maple leaves to make a mini landscape, as you might see in the wild. Even without flowers, this makes something good to look at on our doorstep over winter.

Bulb lasagnes in deep pots are another excellent way of using reticulatas: the iris bulbs make up the top layer, with two or three layers of tulips below (see pages 339 and 365). We use early-flowering crocus in the same way at Perch Hill.

Opposite A square bulb tray with *Iris reticulata* 'Blue Note' coated with foraged moss and dried oak leaves.
Below left *I. reticulata* 'Happiness'
Below right *I. reticulata* 'Katharine Hodgkin'.

It's difficult to get a truly magnificent bulb show in a pot that has just one layer. To fit enough bulbs for something spectacular, they would need to be touching, and like apples on a rack, that's a bad idea. If one bulb starts to rot, the whole lot can be cross-infected. But if you choose the right mix of tulips for the bulb lasagne, you'll get the correct heights of stem to work well together, as well as flowers that come all at the same time. By topping with iris, the whole spectacle gets going in February, just when we all need a little cheer.

Reticulatas are among the cheapest bulbs you can buy, and the outlay is only once, as after everything has gone over, you can easily take the top layer of bulbs out, pull off the dead or dying leaves and store the bulbs in a net bag. Hang the bag somewhere cool in a shed until the following autumn, then plant them again.

There's another bonu with both crocuses and irises: the flowers come first, followed by the foliage. The spiky iris leaves are an excellent foil to the lushness of the flowering tulips. And with crocus, I love their relaxed curve over the edge of the pot, giving a sort of lower leaf storey (see page 100).

February would not be February without these early beauties.

Below One of our tulip lasagnes with *Iris reticulata* leaves contrasting well with the shape of the tulips. Opposite February *I. reticulata* 'Purple Hill' preceding April tulips.

Best of the winter-flowering irises

You can't really go wrong with any Iris unguicularis, which flower the earliest, but there are a few with stand-out class. Next to flower are the reticulatas and histrioides, guaranteed for February in Sussex, and lastly the snake's head iris (Iris tuberosa), from March into April.

1 *Iris histrioides*
'Lady Beatrix Stanley'
With a distinctive scent of violets, this is the cheeriest blue dwarf iris we've grown. It stands out like a beacon; not as velvety as 'Purple Hill', but bright and sparky.

2 *I. reticulata* 'Fabiola'
The bluest of the reticulatas. It is the favourite of our head gardener, Josie Lewis. 'Harmony' is the standard blue, but Josie loves the white on the falls of 'Fabiola', which make it look a bit classier, fresher and, she says, in keeping with the season.

3 *I. reticulata* 'Purple Hill'
With a velvety texture of 'George' (see page 31) and the perfume of 'Scent Sational', this is a winner. It's much taller than the others, too.

4 *I. reticulata* 'Scent Sational'
A very good petal texture and scent.

5 *I. reticulata*
'Sheila Ann Germaney'
In my view, this is prettier than 'Katharine Hodgkin', which is a similar soft grey-blue. A strong and good grower, as well as a lasting garden or potted bulb.

6 *I. tuberosa*
Commonly known as the snake's head or widow iris, this is one of the most exciting wildflowers I found as a child. Rather mesmerising for its scent, colour and velvetiness.

7 *I. unguicularis* 'Mary Barnard'
The richest and most velvety of the Algerian irises.

8 *I. unguicularis* 'Walter Butt'
A lovely, highly scented, pale mauve form.

Violas

Winter-flowering pansies are the first plants I ever grew as an adult. I went to my local west London garden centre in October and bought a terracotta container, some compost and some plants. I remembered a small-flowered pansy my dad used to grow when I was a child called 'Molly Sanderson', and it had slate-black flowers that looked like pure velvet. The garden centre had something similar – I think it might have been *Viola* 'Sorbet Black Delight' – and the label claimed, unlike 'Molly Sanderson', it was a winter and spring (rather than summer) flower.

The person who served me, now thirty years ago, told me I must deadhead the plants once a week to encourage flowering. Like a good, anxious beginner gardener, I deadheaded almost every day.

My window box did well, flowering its heart out until late spring – and I was hooked. This was a key moment in my gardening life and my first introduction to the concept of cut-and-come-again plants. It has been the foundation stone of almost everything I've done as a gardener and I have those mini pansies – and later, a patch of cosmos (see page 252 in August) – to thank.

A cut-and-come-again characteristic in a plant feels generous, optimistic and uplifting – it's incredible that by deadheading, or even better, liveheading (that is, picking fresh flowers to bring inside), you perpetuate flowering. Rather than ruining the display by picking precious flowers from the garden, with the right varieties, the opposite is the case. You are in effect removing the nascent babies and, programmed to reproduce, the plant creates more and more flowers.

The garden, amazingly, looks better, with flowers encouraged not depleted by your harvest. Your house is full of flowers and so is your garden – who couldn't love that idea?

Since filling that first window box thirty years ago, I've grown tons of different violas – and there are some delicious options. I prefer, on the whole, the small-flowered plants (so called violas), but there are some whopper, rather foppish forms (usually called pansies) in single colours that we also grow every year.

A spring pot for flowers in succession from February to April. First comes *Iris reticulata* 'Harmony', its velvet flowers followed by its leaves piercing through a carpet of *Viola* 'Sorbet Phantom'. Then comes the foliage, and lastly the flowers, of *Narcissus* 'Katie Heath'.

I adore the gentle mix of colours of *Viola* × *wittrockiana* 'Mystique Peach Shades' and the similarly toned 'Aquarelle Flambé Toscana'. There's a pansy development that I thought I'd hate, with rouched and ruffled petals, but in fact their flamboyance is fantastic. These varieties, I'm afraid, have 'Frizzle Sizzle' in their names, but they look good as a cut flower for a small jug or vase.

Of the miniature violas, 'Apricot Antique' is a favourite and I adore 'Tiger Eye Red'. 'Apricot Antique' works well as a pot plant and living table centre. It flowers sporadically early in the year, but it really keeps going; by April, the stems are long enough to make it a beautiful cut flower for a small vase. 'Tiger Eye Red' is as it sounds: like the semi-precious stone, rich and beautiful and rather exotic. I loved this stone as a girl and am naturally drawn to these coppery coloured, ink-patterned flowers. They are brilliant for pressing and sticking onto card for Christmas labels and gift tags.

One of the best winter viola pots I've put together included the perfumed *Viola* 'Sorbet Phantom' paired with the big flowers of *Anemone coronaria* 'Jerusalem Blue'. This was another sort of lasagne, but here the anemone corms were overplanted with the viola plugs. To create it, we sowed 'Sorbet Phantom' seeds early in September and they were ready for planting out just four weeks later. You can get all of these violas to flower pretty much whenever you want, including the end of winter, by simply allowing ten weeks from sowing to flowering. The anemone corms were soaked overnight (that plumps them up and encourages them to grow fast), and then planted 10cm (4in) deep in window boxes, with the viola plugs added over the top.

The result was fantastic. The violas started flowering in mid-November and they were joined by the emerging anemone leaves, bright green and crinkly, rather like English parsley. They formed a great pot-filler between the violas and a good backdrop to the mini purple flowers for many weeks.

Then in early February, the anemone flowers started, one corm giving a good fifteen or twenty velvet saucers over the following ten weeks. The viola is tough and resilient, the anemone (in its original form from the warmer climes of Palestine and Israel) a little less so, but in a cold greenhouse or in the shelter of a building, it does well. I left half of the anemone flowers where they were and picked the rest for a vase every week. It was a perfect pairing.

The winner of our greenhouse winter window box trial, with *Anemone coronaria* 'Jerusalem Blue' and *Viola* 'Sorbet Phantom' (see p41 for planting instructions).

Best of the violas

From our trials, it seems almost every viola or pansy sown in early autumn will flower a bit through winter, some more than others. I admit they flower only lightly in the darkest and wettest months (unless in a greenhouse), but even their light flush of flowers is worth having until they hit their stride in spring. These are my rich, glossy favourites.

1 *Viola* 'Antique Shades'
In the loveliest mix of coppery browns and purples, this has superseded the classic *Viola tricolor* for me amongst the small-flowered pansies. One of our favourite edible flowers for decorating puddings and salads at Perch Hill.

2 *V.* 'Apricot Antique'
A lovely warm, soft mix of colours, and flowers with surprisingly long stems, making this good for cutting.

3 *V.* 'Aquarelle Flambé Toscana'
Large, soft, curvy flowers in a mix of pale apricot and coffee shades, with enough crimson markings and central splotches to stop them becoming sickly sweet. They flower a little in January and February, but get going properly in spring.

4 *V.* 'Green Goddess'
Better when sown in early spring for spring and summer flowering, with flowers in gold to olive green.

5 *V.* 'Red Blotch'
An incredibly reliable, long-flowering and prolific viola in rain, sun or even snow.

6 *V.* 'Sorbet Phantom'
Some scent, velvet-textured flowers and a hugely long flowering season, which can include all winter, depending on when you sow them.

7 *V.* 'Tiger Eye Red'
Rich and gorgeous, like the semi-precious stone. If I had to choose just one, it would probably be this. From an early autumn sowing the previous year, the best flowers come from April to June, but they will flower sporadically well before that, depending on the weather.

Practical
January & February

If you like the idea of a jam-packed, flower-filled garden, there are a few jobs that are best done in the first two months of the year.

We do much of our overall garden planning and designing now, working out good combinations for ordering, sowing and growing. We also get in any bare root plants and do our rose pruning and training. We're usually too busy in the late autumn with final bulb planting to fit in much pruning then.

We also start some early sowing. We usually sow sweet peas in November and more at this time of year, just after Christmas. In February, we sow the seeds that take time to germinate, like antirrhinums, and the seeds of plants and climbers (like *Cobaea scandens*), which need to get to a good size before flowering. If you leave cobaea too late, it'll be cut down by autumn frosts just as it starts to bloom.

Planning

January is the month for planning. It's an ideal time to trawl through seed and plant catalogues. We have a Dutch wood-burner in our kitchen, and the first thing I do before a winter planning session is to get the fire roaring. But if I can't have that, then plenty of candles in a range of colours creates a bit of winter cheer.

I dig out all the catalogues we've been sent in the post and settle down with sticky notes (multi-coloured, naturally) and enjoy bookmarking the pages. I still prefer looking through printed plant and bulb catalogues, rather than selecting online. We always grow our stalwarts, all the things mentioned in this book, but I love to add plenty of new things we haven't previously tried.

To get the most interesting range, we never buy from just one supplier, but spread our net as wide as we can, drawing on the information we've learned from our trials the year before, as well as from my visits to trial fields in Britain and Europe. That makes every year exciting – the anticipation of watching these newcomers perform.

For inspiration, I try to find photographs of the garden we've taken through the seasons – it's easy to forget good combinations if you don't keep records and I find photographs more compelling than pages of garden notes.

I also print images I've posted or seen on Instagram (a great week-by-week record), or pictures I've stored of plants I've liked the look of in other people's gardens. I cut and stick stuff together onto huge bits of paper that I pin to the wall, playing around until I feel I've got plant mixes that excite, as well as those that bring a bit of calm.

To really achieve seamless succession, I look back at notes we made during the previous year. I identify the gaps and any areas that were particularly lacklustre, and then I make a black and white drawing of the shape and size of those areas. Then with different sheets of tracing paper (or kitchen baking paper) overlaying it, I add ideas for one season, then the next, and so on. Then we know exactly what to sow and grow.

As with making plant selections, I much prefer doing all this planning on paper, rather than on screen. With a physical overview of the seasons to come, I get a better idea of the overall mix of colours and general feel.

Planting Bare Root Roses

This is the best time to plant bare root roses as they can be safely lifted while they're in the dormant period. Bare root is also the best way to buy a good range of roses without spending an arm and a leg – they are about two thirds the price of potted roses. For my selection, see page 168 in June.

If the weather is mild and the ground isn't frozen or waterlogged, they can be planted out straight away. If it's frosty, put them all in a big pot, cover the roots with compost and keep the compost damp and cool, but try to get them in the ground as soon as possible. Most roses look best planted in groups of three of the same variety, and that's how we plant almost all of our shrub roses here.

* If you have poor soil, improve the whole planting area by incorporating compost or farmyard manure a few weeks before planting.
* Soak the root in a bucket of water for an hour or so.
* Dig a hole that is deep and wide enough to take the rose roots.
* Fork the base of the hole over well to break up the soil, then mix in a spade of well-rotted farmyard manure or compost.
* Water the hole – this will ensure that the roots come into contact with damp soil when planted. If you only water the rose after it has been planted, you risk most of the water simply running away.
* Temporarily place the rose in the centre and lay a bamboo cane across the top of the hole to make sure that the 'union' of the rose (that is, the point at which the rootstock meets the graft, which looks like a knee) is slightly below soil level. If it isn't, dig the hole deeper. This is crucial: if the union is above soil level, you promote the formation of suckers from the root material and these may then outgrow the grafted rose on top.
* Re-dip the roots in water and hold the rose over the hole while scattering mycorrhizal fungi generously – a couple of tablespoons – over the roots. Any that doesn't stick to the damp roots then falls into the planting hole. The mycorrhizal fungi help to establish the rose more quickly. We have done trials of this, and with certain plants like roses, it makes a noticeable difference to growth and flowering in the first year.
* Fill in the hole and firm down the soil with your heel. Then mulch the area well.
* Allow the rose to settle in for a few months before underplanting with plants such as salvias. You don't want too much competing with the newly planted roses for water and nutrients. Continue to water them throughout summer to ensure that they establish well, particularly if the weather is dry.

Below top *Rosa* 'Ispahan' trained onto
hazel benders in March.
Below bottom The same *R.* 'Ispahan'
in full bloom, with flowers from the
ground up.

Pruning and Training Roses

I was taught this method of rose pruning by
the gardeners at Sissinghurst, who follow a
rose training technique that has been passed
down from one gardener to the next since head
gardener Jack Vass started at Sissinghurst in 1939.

By pruning and training the Sissinghurst
way, we get roses covered in flowers, from

soil level to the apex of their dome or cylinder. It gives us the very best rose performance.

If you put every stem of your rose under pressure, bending it and stressing it, it flowers better. Horizontal training overcomes apical dominance (where the main stem is dominant over the side stems), and encourages the side shoots to believe they have an equal chance of flowering – so they do.

We try to prune our climbers and ramblers first, but we don't have many of those, so move quickly on to the shrub roses. All the pruning and training needs to be completed before the roses come into leaf to avoid damaging the leaf buds and shoots as we manipulate them.

For our shrubs, we need to have ready hazel branches of various thicknesses. These can be bought or harvested – they need to be cut in winter before the catkins emerge, so the branches are pliable but bare. The structures we make from the hazel can be left in place for a couple of years.

Climbers

- ❧ Release the climbing rose from its bindings. Remove any old, diseased or damaged wood.
- ❧ When deciding which stems to cut out, always select the oldest first, as they're more likely to carry pests and diseases.
- ❧ Cut some of the previous year's growth down to the base. This encourages new growth to flower and keeps the main stem framework clear and uncongested, helping to protect against fungal diseases and prevent the rose from getting too woody.
- ❧ Prune any laterals coming off the main stem down to 3 or 4 buds. Be brave, if you think the stems still look congested, cut them out.
- ❧ Re-attach the rose stems that remain to the wall, stem by stem. Start with the outermost branches and work towards the middle. Bend down the tip of each pruned stem to form a fan shape and attach to the stem below. Some are pliable and easy to train, others take coaxing.

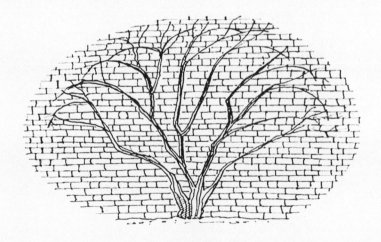

Ramblers

These are generally much bigger than climbers and best left to their own devices, but old wood needs to be taken out to encourage new growth.

Shrubs

Depending on their habit, shrub roses at Perch Hill are trained annually in one of three ways.

For the big, leggy shrub roses that put out great pliable arms that are easy to train (such as 'The Simple Life' and 'Ispahan'), you should:

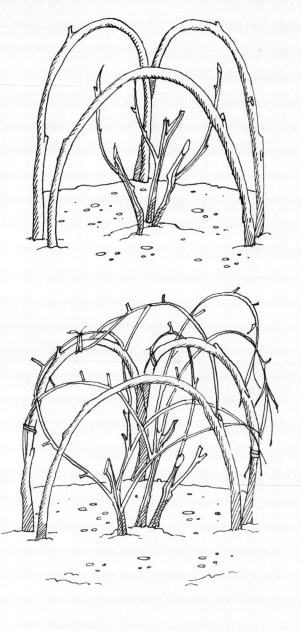

- ✤ Remove all the old, diseased and damaged wood.
- ✤ Make hoops using lengths of hazel with a diameter of about 2cm (1in). Create a triangle of hoops around a single rose or clump of roses (the illustration pictured top right shows the bush without its stems for clarity).
- ✤ Stem by stem, bend last year's wood down and tie it onto the hazel hoops. Start at the outside and tie that in first, then move towards the middle, using the plant's own stems to build up a web shape, tying one stem to the stem below (pictured, bottom right). In the case of large and vigorous varieties, you can create fantastic height, one layer of stems domed and attached to the one below.
- ✤ Without any sign of a flower, this looks magnificent, and in a few months, each stem (under stress) will flower abundantly and will do so on stems curved almost to ground level. It's a miraculous thing.

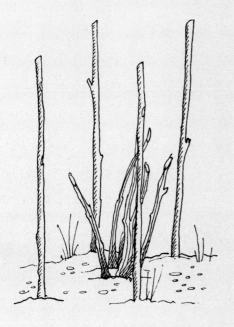

For the tall, rangy bushes with stiffer branches, such as 'Warm Welcome', you should:

❋ Remove all the old, diseased and damaged wood.
❋ Use four lengths of thick hazel (about the diameter of a thumb), to make a cylinder that's hammered into the soil (pictured top left, positioned around a very young bush).
❋ This structure surrounds the rose root clump and all remaining stems can be twirled up the frame.
❋ Finally, when the rose has grown a little, using the same bending technique as the climbers on page 50, every pruned tip is curved and attached onto a length below. To do this, bend down the tip of each pruned stem to form a fan shape and attach to the stem below (pictured, bottom left).

For petite, contained, well-behaved varieties that don't need support, such as rose 'De Resht', as well as small shrubs that produce growth that's too stiff to bend, like 'Darcey Bussell', you should:

❋ Remove all the old, diseased and damaged wood.
❋ Prune the rest back hard to an outward-facing bud, removing at least half the existing length of stems.

Sowing Sweet Peas

To achieve the longest stem lengths, get going with your sweet pea sowing now. At Perch Hill, we also sow sweet peas in November and often right through to March. For my selection of the best sweet peas, see page 206.

❧ I never bother to soak sweet pea seeds. They'll germinate without soaking within 10-14 days.

❧ Sow two seeds to a small pot. We use root trainers, but you can use loo rolls, as both give a nice deep, narrow root run, which is just what sweet peas need. When the sweet pea seed first germinates, it puts down one long root. This breaks off when it hits the air at the bottom of the pot, and like pinching out the tip, the root then throws out lots of side roots further up. When these side roots reach the side walls of the root trainer, they slot into a groove-like channel and are directed straight to the bottom of the pot. They then break off and produce their own side roots, so you get a virtuous cycle of root development, with a root system forming very quickly. If you sow into a short, stumpy pot, the initial root will be shorter and will branch out less. Essentially, a longer root means more root branches, faster, which equals a bigger plant.

❧ Use a loam-less compost with an open structure (such as a multi-purpose potting compost). The plant roots need air, as well as moisture and nutrients.

❧ Dampen the surface and then, with your finger, push each seed in about 2cm (1in) below the surface of the compost. Water again. Label.

❧ Put in a cold greenhouse or cold frame. Don't mollycoddle them – that's the commonest mistake. No heat is key. Heat can inhibit germination, and with it, you're likely to get more seed rot. So put the trays somewhere cold – they're frost tolerant to about −10°C (14°F). A bit of frost seems to do them good.

❧ Guard against mice and voles. Both love sweet pea seed and your whole crop may disappear in one go. If you have the trays in a cold greenhouse, put them on a raised sheet of wood or ply (not hardboard), and make sure that there's a good overhang from the side of the bench so that the mice/voles can't climb up on to the plants. Or soak the seeds in liquid seaweed fertiliser overnight to make them unpalatable.

❧ After a week or so, check for germination every day.

❧ Don't water until you see seedlings start to come through, usually in 10-14 days.

❧ Once the seedlings appear, keep them cool at about 5°C (41°F). This promotes root growth, rather than stem growth, which is exactly what you want. A cold greenhouse or cold frame is ideal. When I started growing sweet peas, I just used a couple of straw bales with a reclaimed window over

Sweet pea seedlings in a root trainer,
showing the vertical root run, which speeds
up root formation.

the top. This keeps the worst of the wet off
and sweet peas don't need warmth.

❧ When there are three or four pairs of leaves,
pinch out the leader – cut or squeeze off the
growing tip between your finger and thumb.
This promotes vigorous side shoot formation,
with the energy of the plant directed toward
growing out, not up.

❧ Every week, check your plants. Water them
lightly if they are dry. If they have started to
shoot again, pinch out any spindly tips.

❧ About a month from germination, check
the bottom of the pot for white roots. As soon

as there are roots visible through the holes,
pot the plants up in their pairs. Don't let them
get pot-bound, they will never be quite the same
again. A slim, deep, 1 litre pot is ideal. Use a
good compost and water them in.

❧ See page 85 in March for planting out
sweet peas.

Sowing Cobaea

Cobaea scandens, known as the cup and saucer vine,
needs to be sown early in the year. It won't flower
until it reaches about 2 metres. Sown late, it tends
to reach this height just when we get our first hard
frost of the autumn, meaning you'll never see
flowers. Sown early, it will fill your autumn with
cups and saucers.

❧ The seeds have a large surface area and are
a wafer-thin saucer shape: sow them vertically
(any end first), rather than flat, into their own
individual small pots (see page 83 for how).

❧ Put them in a propagator (see page 79 to
make your own). They'll germinate, warm
and moist, within 2-3 weeks.

❧ Once the seedlings start to grow, after
about 3-4 weeks, don't leave them sprawling –
they need a climbing frame to clamber over.
Pot them on, and in their new pot, create a
frame for them. We make ours from silver
birch pea sticks, which we weave into mini,
pot-sized teepees (see page 88).

Sowing Antirrhinum

Start sowing the slow-growing annual snapdragons undercover. They are grown as half-hardy annuals, but strictly speaking, they are tender perennials. They have a long growing season and can take up to twenty weeks from seed to flower. Get going with single-coloured varieties such as 'Liberty Classic Crimson' or 'White Giant'. Or try the newly bred Chantilly series, which have a lovely fruity scent, or 'Apple Blossom', which smells of cinnamon. Seeds sown now should flower from June.

Antirrhinums are able to put up with some cold nights, so we often do a second sowing in June to keep them flowering until almost Christmas. We may even sow again in October, storing in cold frames over winter, and planting them out in late March or early April for very early flowers.

- ❀ Antirrhinums have tiny, dust-like seed, so are too fiddly to sow into guttering or Jiffy pellets. The only thing to sow them into is a seed tray (see page 80 for how).
- ❀ They germinate best in moist compost with a little basal heat, so use a propagator or heated mat (see page 79) and leave them in a greenhouse or on a cool windowsill.
- ❀ In a month or so, these seedlings will need pricking out (see page 81).

March

With the sun starting to make itself felt and the days lengthening, March is when I feel my spirits lifting. I, and it seems many gardeners, have a heightened sensitivity to light. Maybe that's why we've chosen to spend our lives gardening.

I like winter hibernation, but even in the south of England it goes on too long for me. November and December are fine, as is the quiet withdrawal of January, but in February, with the drear continuing, I sometimes find it hard. March 1st, and even better the Spring Equinox towards the end of the month, feels good. The banks on the lane and edges of the woods around us start to fill with primroses and the bluebell leaf spears pierce through and lengthen week by week, while the goat willows flare like lightbulbs amongst the silver-browns of all the other trees.

Changes in the garden seem slow at first, then everything rockets. With growth curves steepening day by day, March is the month of mass propagation and getting plants going. It's the peak time for sowing seeds, potting up dahlia tubers and planting out hardy annuals such as sweet peas.

They're usually our first seedlings to go out, with honeywort (*Cerinthe major* 'Purpurascens'), calendulas and *Euphorbia oblongata* close behind. From an early autumn sowing, we plant half of this trio of plants six weeks after sowing, and the other half now, in March. These have been protected in a cold frame in case we have a severe winter. Cerinthe often starts to flower now, particularly if the winter has been mild; if we're lucky, we'll have euphorbia too.

We grow lots of different euphorbias in the garden here and it's in March in particular that we really appreciate them. We have an understorey of the carpet-forming *E. epithymoides* in several places (it's excellent bordering areas planted with tulips), which is invaluable at this time of year for its brightness and lovely again in the autumn when it turns a great mix of ochres and reds. The drive is lined with a self-seeding, self-perpetuating hedge of *E. characias* 'John Tomlinson', with its cheerleader pompoms. I love this evergreen, grey-green foliage plant, which I grew up with in my parents' garden; it starts flowering properly now, its heads rolling upwards as they unfurl.

Snowdrops are merging into carpets rather than splotches. The variety we grow the most is 'S. Arnott', with its whopper flowers, but it arrives a little later than most of the snowdrops and is at its best here in March. Tall, with nodding flowers, it is wonderful picked for a vase.

Crocuses are important too, especially the scented varieties such as 'Cream Beauty', 'Blue Pearl' and 'Snow Bunting'. And I love 'Spring Beauty' and 'Gipsy Girl', whose outer petals look like they've been painted. I cut all of these for small jugs, egg cups and bowls, mixing them with yellow winter aconites. In the warmth and light inside, they open right out like small waterlilies.

Below *Crocus* 'Spring Beauty'.
Opposite *Euphorbia characias* 'John Tomlinson' edging the drive and our south-facing bank where we grow salads and veg.

The fine-flowered hyacinth
'Anastasia', my favourite.

There are, thank goodness, a few early tulips, but above all, narcissi are this month's optimistic flower family, the first to truly fill the garden and give us huge armfuls of stems to bring inside. We can have colour in January and February, but on the whole, it's on a miniature scale; come the narcissi in March, our garden starts its colourful, flowery performance in earnest.

My favourites are the delicate, close-to-the wild species *Narcissus* 'Segovia' and the new, unflatteringly named 'Xit' – both are small, not in any way overblown and as cool as a ballerina. I'm always on the lookout for multi-headed varieties, which are good for cutting and have a better than usual vase life. They're guaranteed mood-enhancers, particularly when scented. In March, we have 'Avalanche', which can throw up stems 60cm tall – it's been in the same place for twenty years and the bulbs must be enormous. 'Geranium' flowers now too, with one of the best scents in the plant world.

As colour in March is still quite minimal, fragrance is crucial. Daphnes and sarcococcas give plenty, and hyacinths of course, which are flowering in the garden. I love the delicate, bluebell lookalike hyacinth 'Anastasia' best of all. These all give scent in spades, but Perch Hill would be a lesser place without its beautiful narcissi.

Opposite A miniature vase of March flowers including crocuses 'Advance', 'Gipsy Girl' and 'Zwanenburg Bronze', and primula 'Gold Lace'.
Below Narcissus 'Actaea' with tulip 'Shogun' along the drive.

Tulip 'Apricot Delight' with a bunch of scented narcissi: 'Bell Song', 'Actaea', 'Pink Charm' and 'Sweet Smiles'.

Narcissi

Adam and I got married on December 31st some thirty years ago and had a party in my mother's house on the west coast of Scotland. We decorated every table, fireplace and window ledge with deliciously scented narcissi, the only British flower grown outdoors and available cheaply in the depths of winter.

The flowers arrived in long white boxes on the overnight train to Fort William, via Penzance and London, from the Isles of Scilly. We put them in old Victorian marmalade jars with a few lichen-encrusted branches. As you moved from one room to the next, you walked in and out of pockets of intensely sweet scent. I've loved paperwhites (*Narcissus papyraceus*) and N. 'Grand Soleil d'Or' ever since.

It was our wedding that spurred me on to choose highly scented, multi-headed, good-vase-life narcissi as the first group of plants to trial at Perch Hill in the late 1990s.

Buying almost any plant from a picture (in a catalogue or online) is tricky. You don't get a feel for each plant's scale and habit, and obviously no idea of scent. With a genus like *Narcissus*, there are so many species and varieties, how on earth do you know which to choose? That's why trialling seemed essential – the same goes for tulips, sweet peas, roses and dahlias.

With narcissi, I knew from a visit to the Isles of Scilly that there were plenty of varieties which lasted not just two or three days once cut, but more like a week if kept cool, but I wanted to know which were the prettiest and most deliciously perfumed (some can be a little camphory).

We sourced the bulbs mainly from Cornwall and Scilly, where there's been a history of narcissi breeding and flower growing for 150 years. The islands at the far south-western corner of England have the ideal climate for these bulbs. Winter and spring are very mild – the mean spring temperature is the same as Barcelona, rarely dropping below 10°C (50°F), even at night – and the summers are usually hot, dry and bright. It's the ideal climate for initiating flower formation in the bulb.

Narcissi harvest on St. Agnes, Isles of Scilly, including 'Martinette', 'Avalanche' and paperwhite 'Inbal'.

The main ones I wanted to trial were the Tazetta types. Unlike other narcissi, that have stems with just one flower that lasts for two or three weeks, the Tazetta group is bunch-flowering, with up to ten flowers on one stem. They also have the ideal cropping pattern: light and over a long period of up to three months. You can pick a crop one week and then again a month later and again a month after that. They are also scented – that's crucial.

I wanted the flowers and leaves of the varieties I chose to be relatively discreet and elegant, so that they would look good in a border, as well as en masse in grass (see page 144). And I also knew which varieties I didn't want in the trial: as a general rule, I'm not keen on the chunky Trumpet group – they tend to have great strips of almost leek-like leaves that look horrid by late spring, and vast cup-and-saucer flowers that seem double a daffodil's natural size.

The narcissi I selected needed to be outdoor, hardy varieties, so I sadly had to exclude the ones we had at our wedding, which would not flourish in most of the UK. However, if the variety could be forced indoors and planted out in the garden once in flower, they made the cut. My final aim was to have a succession of narcissi flowering from the beginning of March until the middle and even the end of spring.

Opposite Before the tulips properly arrive in March, we fill the Dutch Yard with pots of narcissi, which are grouped around the amelanchier. Left to right: *Narcissus* 'Katie Heath', 'Pink Charm', paperwhite 'Inbal', 'Xit' (front left), 'Prom Dance', 'Stainless', 'Lemon Drops' (front right), 'Kokopelli' and 'Martinette'. Below Window box in the Chelsea shed with puschkinia, *Narcissus* 'Sailboat' and the grape hyacinth 'Jenny Robinson'.

Based on those parameters, I selected twenty narcissi for the trial and the characteristics we scored them on included delicious scent, multi-headed flowers (so you could fill a vase with only a handful of stems), their ability to flower for a long period of time and also to become truly perennial, coming back year after year without needing to be lifted and divided.

We also scored on vase life, testing the narcissi in vases filled with plain water, as well as some filled with water and a teaspoon of thick bleach. We also placed some of them in cool, dark locations and others in bright, warm spots. The results here were clear: the flowers in the cool space lasted two to three days longer than those in the sunny one. I always try to put vases outside the back door when I go to bed to maximise any flower's vase life. They last better in the cool. It's worth knowing, that unlike many cut flowers, bleach appears to make very little difference to the vase life of a narcissi. For most flowers, bleach or clear vinegar (which I now prefer to use) kills the bacteria that generate slime, and it's the slime that blocks the stem ends, preventing the flowers from taking up water – but this process does not seem necessary for narcissi.

Below A selection from 2019's narcissi-for-picking trial. From left to right: 'Blushing Lady' has primrose flowers with a coral trumpet; 'Pink Charm' is white with a pink grapefruit trumpet; and 'Bell Song' and 'Katie Heath' are very similar with softer white and bleached-pink grapefruit colouring. Opposite Our first ever plant trial – highly scented, multi-headed, good-for-cutting narcissi. The bulbs have remained in the same beds for over twenty years. Unseen, hidden by the narcissi, are also dahlias, Dutch iris, alliums and hyacinths (see p77). From front to back: N. 'Trevithian', 'Actaea', 'Avalanche' and 'Geranium'.

Best of the narcissi for cutting

To give a succession of flowers from early March until the end of April, this is the list of my favourite varieties for cutting, in order of flowering. As one variety goes over, the next comes out. For the best narcissi for grass, see page 144 in May.

1 *Narcissus* 'Avalanche' (Tazetta) At its best in the first weeks of March, but often with a welcome first flush in February, 'Avalanche' starts things off. It then flowers longer than any other in this selection – it's in the garden for five weeks at least. It is scented and lasts 10 days in water if picked in bud and kept cool and out of sunlight. You can grow it outside or force it inside for early in the New Year. It is a rangy grower, with tall and abundant production of stems. If forcing it in a pot inside, you can stop it flopping around by growing it up through a nest of silver birch (see small pot support on page 88).

2 N. 'Geranium' (Tazetta) The next top performer to flower is 'Geranium', sometimes called the florist's daffodil. It has lots of pretty orange and ivory flowerheads topping each stem, which give a delicious scent and have a long vase life. It will last a week in water if kept cool and out of bright sunlight. This has the RHS Award of Garden Merit.

3 N. 'Actaea' (Poeticus) Tall, elegant stems exude an incredible exotic scent; there's almost nothing better than a large jug of these in the house. This traditional variety is lovely in a border or grass and came top in the trial for longevity, lasting longer in the vase than any of the newly bred forms. The only slight downside is that it is not multi-headed and only flowers for about three weeks.

4 N. 'Silver Chimes' (Tazetta) Flowering about three weeks later than 'Geranium', 'Silver Chimes' has beautiful silvery white petals around a pale ivory trumpet on quite short stems. This is another huge producer, which you can harvest hard over several weeks. You can pick every bud and flower that you see one day, and five or six days later, lots of new ones will have miraculously appeared. This is Mary Berry's favourite.

5 N. 'Xit' (Small-cupped) Classy, discreet, pretty and with a delicious scent, this new variety is unmissable. Easy to grow in a pot, border or grass.

6 N. poeticus var. recurvus Similar but more delicate than 'Actaea', with petals recurved onto the stem. This smells divine and cuts well. Arranged here with the larger-flowered, N. 'Polar Ice'.

Dahlias, narcissi & layered borders

The narcissi trial was a success and varieties we settled on went into the garden in the autumn of 2001. For a couple of years, we left it at that, the narcissi flowering from March until the end of April, with armfuls of scented flowers for the picking. To use the same space in summer and autumn, I overplanted with annual cutting flowers, such as antirrhinums, cosmos and zinnias.

Then I visited Monet's garden at Giverny and fell for dahlias. I wanted to do a trial of the dark, rich-coloured varieties such as 'Rip City', which I first saw there (there's more on this in September). I bought three plants of 'Rip City' at Giverny and then sourced lots of other dahlias in that colour range, but as planting time approached, I had nowhere for them to go.

The penny dropped one early May evening, as I was starting to tidy up the foliage of the earliest-flowering narcissi: rather than adding annuals over the top, maybe I should try dahlias instead. I was used to layering bulbs in pots, so why not try that in borders too? To make room for the new dahlias we would need to cut down the narcissi foliage to ground level a few weeks earlier than ideal, but I felt it was worth a try.

The experiment was a huge success. Rather than having a whole bed dedicated to scented narcissi and then a few annuals for picking and admiring in the summer, we now had a bed with flowers looking good almost solidly through spring. There was a dip in late spring and early summer, but then as the dahlias came up, colour and flowers from July until November. We'd doubled the performance time.

We added the soft pink dahlia 'Gerrie Hoek', which is reliably the earliest dahlia to flower here, usually by mid-June. Scattered all through, we have also gradually added gladioli, as their spires are a good contrast to the bosomy, undulating domes of dahlias. We tend to contrast the colours of the dahlias with those of the gladioli: the almost black *Gladiolus* 'Espresso' looks great with the contrasting coloured dahlias 'Autumn Orange' and pinky-purple number 'Molly Raven'; and the chartreuse green *G.* 'Evergreen' is lovely with

Our bulb lasagne beds with *Narcissus* 'Trevithian' in the foreground with 'Geranium', 'Actaea' and the last of 'Avalanche' behind. The allium foliage is around the outside.

any of the crimson-black dahlias. We put G. 'Purple Flora' together with oranges such as the dahlia 'Happy Halloween', and finally, the pure white and deeply classy and scented acidanthera, G. murielae with gentle-coloured dahlias like 'Gerrie Hoek'.

Over the years, we've pushed the experiment even further. Five years ago, we added the late-flowering Narcissus poeticus var. recurvus and N. 'Rose of May', and planted hyacinth and alliums around the edges. Lined up at the edge of every bed is the safest position for them: if they went in the middle of the bed, their chunky foliage would be in danger of forming a dense leafy canopy that stifles the emerging dahlias.

In November and December, all the beds are mulched deeply to protect the dahlias from frost, with 15cm (6in) of compost dropped over every dahlia crown. We've found that the mulch protects the glads too, which have now perennialised.

In 2019, we added yet another layer, between the dahlias and narcissi, this time it was the rather unfashionable Dutch iris. This flowers exactly when we need a filler, meaning we can maintain a cropping pattern starting with narcissi ready to pick through spring, dahlias from mid-summer and the Dutch irises stepping in during May and June, when only the alliums are up. There are interesting and glamorous varieties such as 'Red Ember', 'Lion King' and 'Tigereye' now available, and they form the perfect third layer. They are true perennial bulbs, too, so won't ever need replanting.

Even if you don't want to devote a whole bed to this idea, you can use the layering in mixed flower borders for an equally brilliant effect. Twenty years on, I could not recommend this bulb planting system more passionately. We started off with just one 6m x 1.2m (20 x 4ft) rectangular bed in the cutting garden, but have added more and more, and now have four, with more planned. They are jam-packed with flower, colour and scent for nine months of the year. Who could ask for more?

Bulb Flowering Pattern

January
Narcissi foliage starts to emerge and you might even be able to pick the odd stem of 'Avalanche'.

February
N. 'Avalanche' usually coming up.

March
N. 'Avalanche', N. 'Geranium', plus the hyacinths.

April
N. 'Silver Chimes', N. 'Actaea' and N. 'Geranium', plus the hyacinths.

May
N. poeticus var. recurvus and N. 'Rose of May', plus Dutch irises and Allium hollandicum 'Purple Sensation'.

June
Dutch irises and first dahlia 'Gerrie Hoek', plus Allium cristophii.

July
All the dahlias, plus Allium sphaerocephalon.

August
All the dahlias, plus the gladioli.

September
All the dahlias, plus the gladioli.

October
All the dahlias.

November
End of the dahlias but we still have a few until we cut them back.

December
After a colourful year, it's time to cut down and mulch the beds for winter.

1 Creating a bulb lasagne starts in October, with the narcissi going in on a bed of grit. Around the bed's edge we plant allium and hyacinth bulbs. These are covered with a soil and grit mix, then a layer of Dutch iris, and lastly potted dahlias.

2 Once the dahlias have died back in the frost, each one is mulched deeply over the crown.

3 In late winter, the narcissi foliage emerges, followed by the flowers.

4 The dahlia foliage emerges supported by a hazel frame.

5 By July, the dahlias are in full flower and harvest until October.

Practical
March

March is one of our busiest months in the garden. It's full-on sowing time for some hardy annuals and huge numbers of half-hardy annuals. We start by sowing the slower-growing hardies, such as *Scabiosa atropurpurea* and *Salvia viridis* blue-flowered, and then move on to some half-hardies such as rudbeckias and tithonia, finishing with cosmos, which germinates and grows super-fast.

It's also the month we try to get all our new dahlia tubers potted. They don't tend to start growing until mid- to late-March here, but they are shooting strongly by April, so we want to get them growing before then.

We also start planting some things out in March. On our heavy clay soil, we focus on planting bare root roses, shrubs and perennials in January and February, but by March, we can plant out sweet peas and other hardy annuals, such as *Ammi majus* and *Visnaga daucoides*, larkspur (*Consolida*) and greater quaking grass (*Briza maxima*). They're tough enough to be outside here in Sussex.

Making a propagator

Our propagator is homemade and you can create something similar with these layers.

1. A sheet of black plastic, or use an empty compost bag that's been split open. This is for the germinating stage, with the plastic keeping moisture and warmth in, but light out, which is the ideal germinating environment for most seeds (warm, moist and dark). A few seeds need light to germinate, but it will tell you on the packet, so don't add this layer for those
2. Seed trays or equivalent
3. Capillary matting (this holds water well and cuts down the need for endless watering)
4. Horticultural electric blanket (to provide basal heat)
5. Polystyrene insulation tile
6. Wooden bench

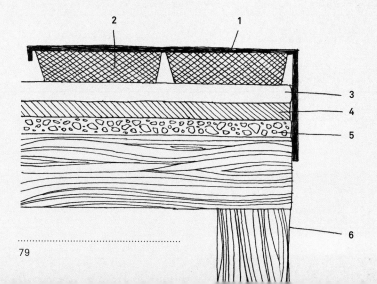

Seedlings spaced widely, so they can be planted straight out from the seed tray.

Seeds, Systems and Labels

We use a different planting system depending on the type of seed. Whichever system you use, always label your sowings. A good tip is to start writing at the blunt end of the label, so you don't have to keep taking the labels out of the pots to read them.

* For tiny seeds, such as antirrhinums, foxgloves, Iceland poppies and nicotianas, sow into conventional seed trays.
* For medium-sized seeds, such as cosmos, rudbeckias, tithonia, as well as grasses such as *Panicum* 'Frosted Explosion', sow into modules or Jiffy-7 coir pellets.
* For large seeds, such as calendulas, cerinthe and zinnias, sow into lengths of guttering – it's the quickest and simplest system for these easy-to-handle seeds.
* For whopper seeds, such as sunflowers and cobaea, sow into individual pots with one or two seeds in each. Some seeds are expensive, so plant one to a pot to avoid any waste.
* For sweet peas, we sow into root trainers. As legumes, these suit a long, deep root run (see page 53).

Sowing in Seed Trays

We do very little traditional sowing into small rectangular seed trays now. With all that pricking out and potting on, it's too much of a palaver. We only use this method for tiny (and therefore difficult-to-sow) seeds, including antirrhinums (which we sow in February, see page 55) and nicotianas, sown in early March.

* Fill the tray with peat-free compost, breaking up any lumps as you go.
* Water the trays before, not after sowing. This avoids displacing tiny seeds to the edges.
* Sow only a small pinch, not a palm full. Treat the seed like gold dust. Sowing as thinly as possible is the route to success.
* If very tiny, sow the seed from a height.

Tiny seeds (like antirrhinums) cannot be individually sown so will need pricking out at seedling stage. Handle by the leaves not stem.

As you sow, move very quickly across the tray. Both these tips will help give you well-spaced seeds over a large surface area.

❧ Cover with a dusting of compost and label.

Pricking Out from Seed Trays

Antirrhinums sown in February and nicotianas sown early in March will need pricking out once they have two to three pairs of true leaves.

❧ Prepare an individual 9cm (4in) pot with peat-free potting compost. Use a dibber or, even better, a pencil to create a small dimple for the seedling.

❧ To lift out each seedling, get as much root as you can by pushing a dibber (or you can use a rigid label) right down to the base of the tray – lift the seedling from there. Avoid touching the stem and lift the seedling out by the leaves. You may need to tear the roots of one seedling from another a little, but don't worry, as long as you're left with some, the seedlings will be fine. You can be firm not tentative. People watching me prick out are always surprised how 'rough' I'm being.

❧ Firm it into its new pot and water. Label with legible handwriting – you'll need to read that same label in several months' time. It's worth keeping a sharp pencil where you sow for this reason.

❧ Place the pot back on the heated propagator until it's time to harden off.

Sowing in Modules

For medium-sized seeds, such as cosmos, use a modular cell system (such as a tray divided into small cells) or Jiffy-7 coir pellets. The large Jiffy-7 coir pellets (around 4½cm/1½in diameter) are ideal. When planted in pellets of this size, the seedlings won't need to be potted on before planting out in the garden – the plants will be big enough to survive as they are.

Sowing cosmos seeds: add one or two
seeds to each coir Jiffy cell.

- With most things, you don't need light at the first germination stage, so you can cover them with empty compost bags that have been split open to keep the heat in.
- Check morning and night for germination: as soon as there's any sign of green, the nascent seedlings need to be given maximum light in a place that's frost-free.
- If both seeds in the dimple germinate, remove the weakest looking one after the first has been showing for about a week. Just pull it out and chuck. This avoids the need to prick out.
- About 5-6 weeks from sowing, usually in May when the risk of frost has passed, plant them out. If you're using Jiffy-7s, make sure you remove the pellet's net in order to allow the plant's roots to run free. It's important you do this, as leaving the net on really holds the plant back.

- The Jiffy-7s need rehydrating in a watertight barrow or plastic crate for 15 minutes before you sow.
- Sow two seeds into the dimple that's at the top of each coir pellet (or fill a modular tray with compost and plant two seeds in each cell). If the seeds you're using are expensive, or there are only a few in the packet, sow just one seed per dimple or cell. Label.
- Place them in a propagator or somewhere warm and cosy; with cosmos, you'll get germination in two to three days.

Sowing in Guttering

We sow large seeds that are easy to handle, such as calendulas and cerinthe, into lengths of guttering. We also use guttering for things such as zinnias, which hate root disturbance (we sow these in April). There's no pricking out or potting on to do, so it's a brilliantly time-saving and very economic system.

- Fill the gutter with a peat-free potting compost.
- Water the compost from the top. Don't bother to drill holes in the bottom of the gutters –

Sowing calendulas into guttering: keep well-spaced so the seedlings can be pushed out individually straight into the garden beds.

Sowing in Pots

For plants with large seeds, sow into small, individual pots. Amongst the flowers, we plant cobaea and sunflowers straight into pots, as they are difficult to get into a Jiffy-7 and their large seeds rot easily in a compact pellet.

* Fill the pots with a peat-free potting compost.
* Water the compost.
* Push the seed in to the depth of about 3cm (1in). Push it in vertically (blunt end down with the sunflower, with cobaea it doesn't matter), rather than flat. The compost usually folds over the seed and buries it, but you can cover with compost if this hasn't happened. Label.
* There's no need to water again straight away. Put on a heated propagator until they germinate.
* Once germinated, water every few days when the compost starts to dry on the surface.
* For hardy annuals, such as sunflowers, harden them off briefly and then plant out into the garden once you see roots appearing in the holes at the bottom of the pot. Keep half-hardy annuals and tender perennials, such as cobaea, protected inside until the chance of frost has passed. You will need to pot on the seedlings as they grow.
* See page 157 in May for how to plant out from pots.

simply leave the ends unblocked, so the water can drain away.

* Place the seeds about 3-4cm (1½in) apart on the top of the compost. Don't push them in until you've laid all your seed out, that way you won't forget where you've placed them.
* Push each seed in lightly with your finger, so it's just under the compost surface. Cover and label.
* Water every few days, when the compost starts to dry on the surface.
* See page 158 in May for how to plant out from gutters.

'If in doubt, pinch out', that's our motto here. This promotes axillary bud formation and a stronger, more flower-productive plant.

Pinching Out

For all seeds you plant, once the seedlings have three pairs of leaves, pinch out the leader of the plants (the leader is the growing tip). Just squeeze it off between your finger and thumb. This promotes vigorous root and side shoot formation, as it removes apical domination, and the energy of the plant can go into growing down and out, rather than up – it makes for a more floriferous and long-living plant.

Planting Dahlia Tubers

You can plant dahlia tubers in a couple of ways, depending on how many you have.

If you only have a few tubers:

* Plant them individually into a 3-litre pot. A 2-litre pot is often too small for bigger tubers and you don't want to force them in and damage the root, and a big pot also allows the tubers to grow happily until the frosts are finished when they can be planted in the garden.
* Plant each tuber in peat-free, multi-purpose potting compost, just under the compost surface, not buried deeply.
* Plant them stem up, with the tubers (which remind me of a bunch of sausages) hanging below. Water well.

* Place on a heated propagator, or somewhere light and frost-free, until they start to shoot (which may not be until the end of April).
* Water when the compost starts to dry on the surface. Don't overwater as this will make them rot.
* Once you see shoots (and roots appearing in the holes at the bottom of the pot), plant out into the garden, but only if the chance of frost has passed.

If you have huge tubers or lots to bring into growth:

* Lay them out in a shallow tray and cover them with moist compost. Water them and keep them protected from frost.
* Once they've started shooting, plant them out individually (spaced about 60cm/2ft apart) in the garden.

In both cases, they'll start to shoot in a few weeks, at which point you can take cuttings (see page 125 in April).

Planting Out Sweet Peas

We did a trial a few years ago in which we planted out a few of the same variety of sweet pea each week from the middle of February to the end of May, and it was the ones planted in the middle of March that did best.

If you sowed your sweet pea seeds in January

or February (see page 53), the seedlings should be romping away now. Once you have got bushy seedlings 5-7cm (2-3in) tall, with roots coming out of the bottom of the root trainer, you need to plant them out. But first, you'll need something to support them. We plant ours around a teepee or a run of pea sticks covered with jute netting – this gives the sweet peas something lovely to climb over. If the birch or hazel wood is harvested in February or early March, the sap is rising, which means the branches will be pliable and not yet in leaf, which is ideal.

Making a teepee

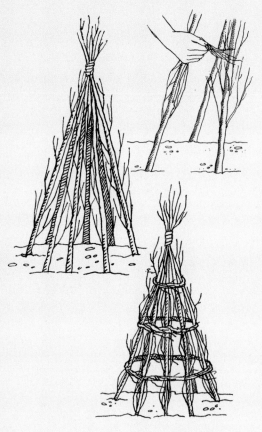

❀ Teepees can be made from bamboo canes, or hazel or silver birch. The supporting canes or branches are referred to as 'uprights', and these need to be at least 2m (6ft) tall. You can supplement these with smaller sticks pushed in between each upright around the base. With bamboo canes, you need to add a network of twine between the uprights to create an efficient climbing frame, which doesn't look so good, but does the job. For hazel and birch, you can use bundles of thinner side branches to wrap around the uprights. The twiggy nature of the thinner birch branches make them the best climbing frame, giving the plants plenty of handholds on which to climb.

❀ Whatever you use, draw a circle, and following that, push eight uprights into the ground, sinking them 20cm (8in) deep. It's key to secure the teepee well into the ground. The circle should be about 1m (3ft) in diameter.

Making a helter-skelter teepee from
silver birch.

* Gather the eight uprights together and tie
with a robust flexi-tie or twine at the top.
* If using birch or hazel, you need to
intertwine the smaller branches together.
Start at one upright in the circle, gather all
the thinner side branches about 45cm (18in)
from the ground and hold them together
in your right hand. Twist horizontally.
Carry on twisting until you get to the next
upright and twist the second bundle, binding
and weaving it in with the next and so on
until you get back to the beginning. Reverse
back on yourself to tie off any loose ends.

* Next, move 45cm (18in) up towards the top
of the teepee and do another layer in the
same way. For tall teepees, you may have
room to add a third layer.
* We also make teepees with one long spiral
from the bottom to the top of the teepee, so
the whole thing looks like a helter-skelter. To
do this keep twisting in a spiral, gathering
as you go, until you reach the top. You'll
probably need to work from a step-ladder.

Making an aisle
* For this, we use hazel. Go for branches
that are 3-4cm (1½in) diameter and about
2m (6ft) tall.
* Uprights are placed at 1m (3ft) intervals
and hammered 30cm (12in) into the ground
– do this from a ladder with a rubber mallet.
* Shorter hazel struts are added to strengthen
the frame at each corner.
* Jute netting is then attached, stretched
taught over the hazel frame.

Planting sweet peas
* Plant two seedlings to each upright – this is
why we sow and grow in pairs (see page 53).
That will give your structure good coverage
and plenty of cut flowers, without the hassle
of cordon training.
* The pair of plants need to be spaced at 15cm
(6in) around the base of a teepee, and 5-7cm
(2-3in) away from the support. If planting in
an aisle, alternate either side of the frame.
* Sweet peas are some of the hungriest and

moisture in the soil and protect against slugs by encouraging blackbirds. If you mulch with your own garden compost, ground feeding birds will come and rootle around in it for worms, slugs and beetles.

❧ If you have a problem with slugs, surround your plants with slug prevention, particularly at the early growing stages. We use a 30cm (12in) wide strip of horticultural grit that is about 3-4cm (1in) deep.

Small Pot Frames

If you're planting tender perennial climbers, such as cobaea or Spanish flag, you'll need to encourage them to grow up a frame as seedlings. Left trailing, both top and root growth slows. We use similar pot frames for the narcissi and other bulbs we force indoors, as well as for larger pots (see page 125).

thirstiest plants we grow, so dig a good, deep hole for each pair of plants, or a 30cm (12in) wide and deep trench for an aisle of plants.

❧ Fill up to about one third with peat-free compost or farmyard manure. The manure acts like a sponge and slowly releases food and water at the sweet pea roots.

❧ Cover over soil to protect the youngest roots from the richness of the manure.

❧ Water the hole and plant the pairs of plants.

❧ Tie them into the base of the teepee or aisle using twine or twist ties.

❧ Mulch. Mulching helps feed the roots, retain

❧ As soon as seedlings reach 10cm (4in), pot them on into deep 1-litre pots. Use a rich, peat-free compost, as they're going to be in these pots for another 6-8 weeks.

❧ Push birch or hazel twigs (about 60cm/ 2ft tall) firmly into the compost around the perimeter of the pot. Gather the ends and tie them in a bundle at the top using twine or a flexi-tie. The climber can then scramble over this twiggy teepee.

❧ Pot on into a larger pot and larger frame for outside after the risk of frost has passed.

Making a small pot frame for February-
sown *Cobaea scandens*.

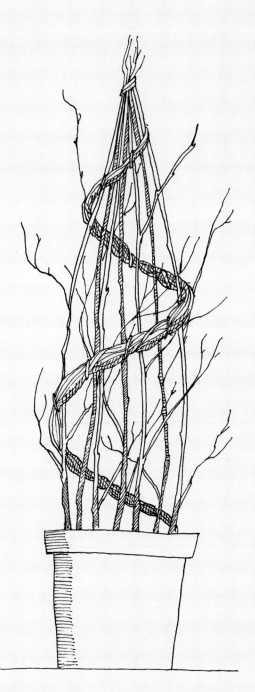

April

Every time I step outside in April, new flowers have emerged. If I go away for a few days, the garden has completely changed by the time I get home – and I love that.

Everything fills out so quickly. In March, there's still plenty of bare soil to be seen, whereas by the middle of April, the whole place is merging into a continuous carpet of spangled colour with anemones, fritillaries, early-flowering biennials (particularly the wallflowers and honesty), and most striking of all, tulips.

We do a tulip trial every year and must have grown nearly five hundred different varieties over the last few decades. Each spring, another thirty or so are tested. That's one of the reasons a walk in the garden in April is a source of excitement: there are always new tulips to assess and hopefully admire.

Tulips definitely give us the April zing, but we need perennials at Perch Hill – and our roses and trees – for the whole place to be truly colourful. I've already mentioned the acid-green feather boas of *Euphorbia characias* 'John Tomlinson' that start up in March, and they have an ever-increasing presence as the weeks go by – a perfect contrast to the richly coloured tulips we grow.

New rose foliage gives important colour throughout the garden and we accentuate it by planting matching tulips. At this time of year, we focus not on the rose's flower colour, but the shades of its stems and leaves. Tulip varieties such as the bronze 'Brown Sugar', coppery 'Request', bi-colour crimson and bronze 'Slawa' and softer brown 'La Belle Époque' all highlight the unfurling leaf colour of the roses. While the red and maroon tulips such as 'Jan Reus', 'Couleur Cardinal' and 'Queen of

Night' pick up on the pink, red and crimson stem tips of almost every rose in April.

There are the crimson shoots of the phlox hybrids, which are up a good six inches, and delphiniums and artichokes are moving so fast they're on the brink of needing staking. We use local wood, mainly silver birch and hazel, for this and are wary of willow here (since it roots so readily on our heavy clay). We don't use bamboo canes either – they look stark in the garden and don't feel quite right.

One of our favourite April jobs is creating support nests and domes to stop many of our perennial plants collapsing when at full height, and this is when we're making teepees to go on top of the huge garden pots (see page 125). These structures stand at about two metres tall on top of the pot, giving immediate vertical spires of nearly three metres. When the frosts are over, in a month or so, the mini 1-litre pot teepees made in March to support the growing climbers (see page 88), are transplanted straight into these large pots, and they quickly romp away. Our tender perennial climbers, such as purple bell vine, black-eyed Susan vine and Spanish flag, are so crucial for colour late in the year, so we want to be ready to plant them out as soon as the chance of frost has passed.

The amelanchier trees in the Dutch Yard are at their fleeting, blossomy best, and our cherries and crab apples are starting to open into cumulonimbus puffy white clouds. The pure white blossoms are, to me, the prettiest, with *Prunus* 'Tai Haku' and *Malus* 'Dartmouth' and *M. hupehensis* my firm favourites.

Almost as lovely as the blossom on the branch is the natural confetti carpet of petals amongst the daises after a night of April winds. The purity of this is the perfect contrast to the bold and brilliant tulip parade, sort of Paris meets Mexico City. I love having both, not quite next to each other, but close by. To me that's April: crisp and classic, intense and bold.

My garden dream is to have the place so full of flowers I can repeatedly cram the house with buckets and vases but, crucially, leave no gaps and holes in the borders. My father instilled in me a sense of guilt about stealing the view, but in April, I can truly bring the outside in without a worry. Every view, every border, is so brimming with flowers and colours that I can really pick armfuls guilt-free.

Tulips and sweet pea teepees in the Annual Cutting Garden.

Tulips

I first fell in love with a tulip thirty years ago in Crete.

My botanist father had written an article for the *Alpine Garden Society Journal* recounting a trip he'd taken there in the early 1950s, when he went in search of the rare *Tulipa saxatilis*, a rich pink tulip with a large egg-yolk yellow splotch at its heart. He had been asked to help decide whether it was a separate species to the almost identical *T. bakeri*, or whether the two were actually one.

His visit was to the White Mountains on the west side of the island. There, in an old stone quarry, he found three *T. saxatilis* in flower, growing in a crack in a boulder. With *T. bakeri* also nearby, he was able to determine that the two had enough differing characteristics to warrant a separate classification.

When I met my husband Adam, we thought it would be fun to retrace my father's trip. In his article there were (surprisingly) clear instructions on where to go and when; Adam was confident he could find the spot. And we did find it, and the boulder and, incredibly, we found the tulips. But sadly the flowers had gone over, the pink petals dropped.

Amazed the tulip was still there forty years on, but also disappointed not to see it in flower, we headed back east towards Heraklion and the airport. Half way there, we stopped at the small town of Spili for the night and went looking for a place to eat. There wasn't much open in early April, with the tourist season yet to begin, but one taverna looked good. In we went to find, not three rare tulips on a boulder, but a jug of thirty sitting on the bar. It was one of the most triumphant moments of my life – and from then on, I've always loved tulips.

The taverna's landlady told us that she had picked the huge bunch from the high meadows above the town, which every spring turns into the most wonderful wild garden of Mediterranean flowers. First come *T. bakeri* and *T. saxatalis* in March into April, followed by *T. doerfleri*, a Cretan endemic, which flowers for Greek Orthodox Easter. I try to go back there as often as I can in spring, as the fields of tulips, mixed with orchids, snake's head (or widow) iris,

The cultivated equivalent to the wild tulips we went to find in Crete: *Tulipa saxatilis* (Bakeri Group) 'Lilac Wonder'.

fragrant daphnes and wild narcissi, create a paradise that's hard to resist.

Our new experiment at Perch Hill is to recreate these fields in miniature by using old sinks raised on brick plinths and filling them with precious and fine alpines, including tulips. It's a traditional way of growing alpines, recreating a different soil type and landscape to that of the garden, so that rare and unusual plants can survive. We haven't got this quite right yet, but what we do have is an impressive – if rather less delicate – swathe of mainly hybrid tulips.

Tulips are pivotal to Perch Hill and we have them everywhere. We have tulips with the roses and in the Oast Garden, brilliantly coloured like boiled sweets and enough to raise anyone's spirits as they walk through. We have tulips in the Dutch Yard that highlight the red brick colour and purple-blue headers from which all our old walls are made. And we have them round the lawn in calm colours as a complement to the green of the grass. We have tulips down the drive that are perennial varieties, there to stay, and we have tulips in almost every colour on our productive edible plant bank. And most pleasingly of all, we have tulips in pots all around the garden,

A picture taken by Adam, my husband, of *Tulipa doerfleri* in the wild gardens around Spili, Crete.

enriching every single inch of the place. They are one of the greatest and cheeriest of flowers.

I aim to pick a bunch of tulips every few days, so we can revel in them inside as well as out. In March, I often pick a whopper vase of one of the Impression series in pink, or a mix of all their colours together – pink, coral and orange. They stand as tall as foxgloves, and just like them, their stems twist and curve in a brilliant and dynamic way. I love the mix of their vast lollipop flowers with their vigorous-looking leaves, so only strip the bottom leaves which will sit under the water line.

At the other extreme, I love a line of single stems of the elegant, pointy-petalled Lily-flowered group, such as 'Ballerina' and the tulip I registered, 'Sarah Raven' – they both flower all through April. They're great down the centre of a table, some left as tall as they come, others cut short so you get a rhythm of ups and downs. I enjoy playing around, pulling together new arrangements from the tulips on trial, sometimes mixing them with the odd stem from the previous year's trial. That's how much of my time is spent in April: experimenting with bunches and working out what I love best in colour as well as form. I always think you only truly know a tulip if you've lived with it and looked at it on your kitchen table for a week.

Sourcing tulips at trial fields

I started our tulip trialling in earnest when I realised that some of the tulips I really loved were disappearing from the market and no longer available to buy. I was always being reassured by the Dutch breeders and bulb suppliers that the replacements would be even better, but again and again, the new varieties weren't a patch on the old.

Ten years ago, it was 'Generaal De Wet', a nice early orange with good scent, that disappeared. Then 'Orange Favourite' went, the most fabulous Parrot tulip with crinkly petals, freesia scent and real glamour that would bloom until mid-May and overlap with the first alliums. It was ideal as the baton carrier from spring into summer. Next to (almost) go was 'Recreado', now hard to come by, it is one of the loveliest purple tulips, made classy by the blackcurrant staining down its stem. And then we had to mourn 'Bruine Wimpel', which flowers in copper, bronze and silver all at the same time and was as good in bud as when its petals curl and prepare to drop four or five weeks later. It was my all-time favourite tulip, and utterly unique.

Pots in the Oast Garden planted with bulb lasagnes of tulips 'Sarah Raven', 'Lasting Love' and 'Ballerina' (see p365). The foliage of *Crocus* 'Flower Record' softens the edge of the pot and hides the plastic inner pot, with the feather-boa flowers of *Euphorbia characias* 'John Tomlinson' behind.

Certainly with 'Bruine Wimpel', its demise was down to a fungus (*Fusarium oxysporum* f. sp. *tulipae*), which rots the bulbs – and I imagine that's probably true of the others as well. Certain varieties seem more prone to diseases, both bacterial and fungal, and are partially or entirely wiped out. But fashions play a part: varieties go out of favour and so are dropped for new ones.

I understood early on that this disappearance was going to be a constant problem for me, as well as for anyone who loves their tulips, so I needed to find my own replacements. It's like your star pupils leaving school: you know you have to bring on the next generation.

So, every spring for the last twenty years or so, I have visited bulb trial fields and Keukenhof gardens in Holland. The Dutch have such a long history with the tulip and have built up a very broad knowledge, investing heavily in research, facilities and technology for decades – and in many places, they also have just the right freely drained soil. Those two things combine to make the Netherlands the centre for tulip breeding and commercial bulking up.

When creating a new tulip, the hybridiser will select two parents with particular desirable characteristics. It might be good scent, good shape, longevity or excellent colour. The pollen from one parent is placed on the stigma of another, and the seed then collected once ripe. This is sown and grown to flowering bulb size, which may take up to five years.

Each individual bulb will produce a different flower to its sibling from the same batch of seed. The hybridiser then selects a few offspring they feel are particularly interesting and nurture only them, discarding the rest. These chosen few are gradually bulked up from just the one bulb in the first flowering year, to perhaps thirty a few years later.

It's at that stage, with a field of many different new forms, that selection panels visit; different groups of people come along at different times and pick out the ones they think are worth investing in, placing a marker by the tulip they are interested in. The rest are binned, never to see the light of day.

Once selected, the individual tulip is then developed, with each mini bulbil (that forms around the base plate of the mother bulb) harvested and grown on to flowering and reproduction size (again, this can take five or six years). And the process keeps going until that particular tulip has bulked up enough to be sold by the thousand or even tens of thousands.

It's at the selection stage that I visit, looking for stand-out tulips in huge fields of hundreds of new varieties. The process of choosing a new hybrid seems a little random and at the mercy of the selector's tastes and aesthetics. In the past, the selection panel (often all male) has tended to steer toward robust and strong-looking tulips, big, bold colours, and lots of bright reds and yellows. That's never been what I look for. I am continually looking for the dark and rich. I'm particularly obsessed by unusual brown tulips, either milky coffee, or even better, chestnut-coloured varieties. And I like height and elegance, rather than chunkiness, and, of course, there's something magical about tulips with scent.

One of my favourite selections was in 2015 when I fell on the tulip now called 'Sarah Raven'. I had been taken to a trial field by two Dutch bulb experts and friends, Carien Van Boxtel and Dicky Schipper. It was a horrendously cold and rainy day; grumpy and longing for a cup of tea, I spotted what looked like an interesting tulip in the middle of a row about thirty metres away. It wasn't hugely showy, but just what I was looking for: a classic lily-flowered shape in a true burgundy-crimson.

Tulip 'Sarah Raven' with a whopper partner, 'Menton', growing through a carpet of fragrant wallflowers.

The Farmhouse Garden with soft, pastel-coloured tulips: 'Ballade Silver', 'Lasergame' and 'Exotic Emperor' in the ground; and 'Weber's Parrot', 'Green Wave', 'White Touch' in the pots with summer snowflake.

There is a tulip called 'Burgundy', but it isn't really burgundy at all: it starts dark, but turns bright purple as the flowers develop, which is always such a disappointment. By contrast, I remember the classiest tulip in my parents' garden when I was a child: it was almost black, like rich, plain chocolate. I remember seeing it – or perhaps something very similar – at Sissinghurst when I first started going there in the 1990s. They'd had it in the garden for years and were unsure of its name – and now it's gone.

The discreet, sophisticated and truly beautiful tulip I found in the field that day was identical to the tulips seared into my memory. No one else on the selection panel had seemed interested and it was about to be discarded. In went my marker – I was overjoyed. We trialled the tulip at Perch Hill for a couple of years and it turned out to be just as good as I had hoped: long-flowering, healthy and pretty perennial. Exactly what you want.

Tulip trials at Perch Hill

We've done many tulip trials here over the years, focusing our research on different characteristics each time.

In our very first year of trialling, we tried out reputedly early

My first proper tulip trial in 2004, which included these rich crimsons and dark oranges.

tulips and then late. It was a priority to have a long-lasting tulip display. When I began gardening at Perch Hill, I ordered tulips just because I liked the look of them in a catalogue, paying no attention to their flowering time. We would have a massive colour bonanza for about three weeks in late April, but not enough before or after that prized period. I quickly realised how choosing varieties carefully can really stretch the flowering season. As most tulips flower mid-season, you'll inevitably have plenty then, so it's worth consciously selecting some that flower early and some late to extend the show.

Not long after, we looked for varieties that reliably flower for a long period. The duration of flowering has become more and more important to me as I've come to grow more tulips in pots than anywhere else. And we pretty much always test for vase life, using a similar system as for the narcissi (see page 66). And every year, we bring in new varieties I've seen in Holland, not testing them for anything in particular, but just to see if they have any outstanding characteristics, particularly scent, which is so valuable in tulips.

To really get the best out of trialling, we grow the plants in a way that's as close as possible to how anyone might grow them in their garden. We plant the bulbs in November (see page 365), and for varieties we want to come back year after year, we feed them once on their way up and once on their way down after they have finished flowering.

Some people advise using a slow-release NPK fertiliser just before the tulips come up, usually around the end of February or early March. But the best feed regime we've found is to use potash (readily available in comfrey tea, see page 193 for the recipe) at the early emergent moment. Then we add a quick-release NPK fertiliser (we use a mix of comfrey tea and nettle tea) after removing the flowers. It's an easy way of sticking to organic, rather than chemical, preparations.

If we want our tulips to be perennial, we also deadhead them: just break or chop the head off (we often use hedging shears in our larger beds) and leave the stems standing. With just the head removed, the plant can continue to photosynthesize and feed the bulb instead of making new seedheads, ideally storing enough starch to flower well the following year and beyond.

By deadheading, you also minimise the risk of spreading the fungal disease tulip fire. The spores of this disease are usually found in the petals, so by removing them from the plant before they drop off, you prevent the spores from falling on the ground where they will survive until the next season.

Tulip groups

The tulip genus is divided into 15 official groups, as well as a few unofficial groupings that are also useful (including French and Lefeber). Knowing which group a tulip belongs to can help you narrow the search to the ones that are right for you.

Species
These are ones collected from the wild, or the hybrids that have been cross-bred from them. They will self-sow and naturalise. There's a whole hybrid range emerging, including 'Annika', one of the best for shallow pots.

Fosteriana
One of the earliest groups to flower, with an unfortunate preponderance of red and yellow. But I love the ivory and elegant 'Purissima'.

Kaufmanniana
The earliest of the earlies, which often have overly large heads to their length of stem, but there are a few notable exceptions, including 'Ice Stick'.

Greigii
Another very early group, which tend to have top-heavy proportions, but 'Für Elise' is a favourite, lovely in pots.

Double Early
The nectaries of tulips in this group have been bred to be secondary petaloids, so they are sterile – not good news for pollinators, but it makes them flower for almost twice as long as average. We grow these with honeywort for the pollinators. The proportions feel wrong to me, with rather large heads on short stems. An exception to this mismatch is 'Chato', which is showy and looks like a peony.

Single Early
These are short, yet elegant tulips in rich colours, often with pretty foliage. Early to flower, usually here in early April.

Darwin Hybrids
These whoppers are the sort of cartoon of a tulip that a child might draw. They are reliable at reappearing every year, but you'll find some reverting to reds and yellows as they age.

Triumph
This group originated from the 19th century, when breeders were looking for a tulip that was suitable for forcing and for growing in the garden. They are called Triumph because of their larger flowers, strong stems and huge colour variation.

Lily-flowered
These are often referred to as the catwalk models of the tulip family. They include the must-have scented 'Ballerina', sumptuous 'Merlot' and my tulip, 'Sarah Raven' (which is not as tall as the rest).

Fringed
The top edge of the petals look as if they've been cut with

crimping shears. I used to have reservations about this lot, but have fallen for 'Louvre Orange'. It's perennial, elegant and the perfect partner for 'Brown Sugar'.

Multi-flowering

Each stem holds three to five individual flowerheads. We grow the blousy 'Belicia'; just three stems will fill a small vase.

Double Late

These have plump, bowl-like flowers with multiple rows of petals. They are sterile (as with Double Early, so similarly long-flowering), but the proportions of stem height and flower size are balanced. 'Antraciet', 'Black Hero' and 'Angélique' are long-standing favourites. These are one of the best for pots and picking because they flower and look good for so long. With their complex, robust structure, they stand up well to wind and rain. Star performers.

Single Late

The late counterpart to Single Early, a good reliable group of tulips that flower from mid-April. We grow lots of 'Caviar' and 'Queen of Night'.

Viridiflora

The tulips with a green flash on their petal. They seem to be very reliable at reappearing. Look to 'Spring Green' and 'Artist'.

French

Famous as cut flowers, these are real whoppers, with very tall stems and a long flowering season well into May. 'Maureen' is a great white. We grow and love 'Avignon' and 'Menton'.

Parrot

Frilly and flamboyant, these are like parrots landing in the garden. Most of them have flowers which include several colours over a main base colour. We find them pretty perennial here, and if I had to choose one, it would be 'Green Wave'.

Lefeber Hybrids

These are among the tallest and latest tulips to bloom and are sports of the incredibly tall 'Temple of Beauty'. We have also grown and loved the stature of 'Elegant Lady' and 'Perestroyka'.

Best of the tulips

Listed in order of flowering

Tulips are in my top five families of garden plants. If you get them in the ground they will flower, it's as simple as that – and they'll give you an incredible procession of shapes and an invaluable range of colour early in the year.

Early-flowering tulips

It's such a boon to have tulips that you know will reliably flower in the middle of March, a good three or four weeks earlier than most. This seems to bring spring forward in the year, which can only be a bonus.

1 *Tulipa turkestanica* (Species) As with many species tulips, this flowers in March and goes over quite quickly. But with this tulip, you get great architectural seedheads that last well, even into June.

2 *T.* 'Ice Stick' (Kaufmanniana) When I first saw this, I wasn't sure, but it has won its way onto my list of favourites – it has unusual grace for a Kaufmanniana variety, is early and long-flowering.

3 *T.* 'Apricot Emperor' (Fosteriana) A softer apricot version of 'Orange Emperor', with nice green markings on the outside of the petals when the buds first come out in mid-March.

4 *T.* 'Purissima' (Fosteriana) With ivory flowers, this is early flowering, long-lasting and super-perennial.

5 *T.* 'Exotic Emperor' (Fosteriana) White and green, huge-headed, semi-double and long-flowering; it has great garden presence.

6 *T.* 'Light and Dreamy' (Darwin Hybrid) Opens soft pink with a mauve wash on its outer petals and deepens as the flowers develop. This, together with many

others in the Darwin Hybrid group, are famously perennial. This one has been returning at Perch Hill for many years.

7 *T.* 'Chato' (Double Early) Cut or growing in a pot, these look like a vase of peonies. Expect flowers to open in early April.

8 Impression series (Darwin Hybrid) These are whoppers with flowers that open up to be the size of single peonies in a range of pinks and reds. I particularly love 'Apricot Impression' (pictured) and 'Salmon Impression'.

9 Delight series (Darwin Hybrid) Similar to the Impression series, but on a more modest scale. I like 'Apricot Delight' (pictured here and on page 64).

Late-flowering tulips

1 *Tulipa* 'Victoria's Secret'
(Parrot)
Huge-headed purple tulip,
which flowers long and late.
Very showy and fairly reliable.

2 & 3 *T.* 'Evergreen' (Triumph)
and T. 'Green Power'
(Viridiflora)
With all-green flowers, these
classy and unique tulips bloom
late and last for ages. 'Green
Power' is newly bred and is
hugely long and late flowering,
into June in some years.

4 *T.* 'Vovos' (Parrot)
This is newly bred and not
a strong grower, so it's still
expensive, but it looks strange
(an unusual mix of mauve and
café au lait) and glamorous.
Flowers in May.

Long-flowering tulips

You'll find that the varieties with lots of green in the petals tend to flower the longest and age elegantly. The green section of the petals is more like a calyx and has a more robust texture than the usual petals.

The dark reds also seem to last better than most, and those in the Double Early group are the same, flowering for a good month and lasting at least two weeks in the vase. T. 'Exotic Emperor', T. 'Spring Green' and others in the Viridiflora group are also appropriate for this category, as is T. 'Evergreen'.

1 *Tulipa* 'Abu Hassan' (Triumph) A deep crimson with golden edges, this tulip looks good for five weeks at least, from lovely buds to dropping petals.

2 *T.* 'Green Wave' (Parrot) A huge, green and pink, wavy-edged Parrot with glamour. It is exceptionally perennial and long-lasting in the vase.

3 *T.* 'Palmyra' (Double Early) All the Doubles flower for ages because they're sterile (see page 108), and 'Palmyra' is the record holder amongst them. Even though it was hot and sunny during our trial, it flowered for five weeks, with 'Copper Image' and 'La Belle Époque' almost as long.

4 *T.* 'Lasting Love' (Triumph) This reminds me of sealing wax in its deep red texture and colour. Its name is apt as it flowers for ages!

5 *T.* 'Black Parrot' (Parrot) Frilly edged, exotic-looking and very nearly black.

Tulips in the Annual Cutting Garden. Front to back: 'Ronaldo', 'Caviar', 'Bigi Brasa' and 'Continental' emerging through a sea of wallflowers planted the previous autumn.

Naturalising and perennial tulips

We all want bulbs to be perennial, so we can plant them one autumn and still be enjoying them for five or ten years to come. For our borders in particular, I am obsessed with tulips that reliably reappear.

In Holland, most gardeners treat tulips as annuals and take them out and bin them at the end of spring, or they lift them. They don't leave tulips in the ground because they fear the risk of tulip blight and spores building up in their soil.

In the UK, most of us feel that binning them is wasteful, and lifting them is too much work, so we leave them where they are planted. We certainly do that at Perch Hill. This habit means that over the years we've

been able to note the tulips that seem to be the best perennials, remaining healthy from one year to the next.

You can often tell if this is occurring as you will see one-year-old, non-flowering foliage tussocks right by an original bulb. This indicates that the bulbs are starting to naturalise. They may be so-called 'blind' for a year or two (meaning they won't flower), but after that, they will bulk up into mini-clumps, a lot like narcissi.

From experience, I can confidently say that those in the Species, Fosteriana, Darwin Hybrid, Viridiflora and Lily-flowered groups are the most reliable. Stalwarts in this category already mentioned in the early and late lists include 'Purissima', 'Exotic Emperor', 'Light and Dreamy' and 'Green Wave'. There are plenty more...

1 *Tulipa* 'Purple Dream' (Lily-flowered)
A boiled-sweet purple with large flowers and really tall stems, this one has the added benefit of being early, opening at the beginning of April.

2 *T.* 'Ballerina' (Lily-flowered)
A brilliant orange with the fantastic scent of freesias. My all-round favourite tulip for borders, pots and picking. It is very special.

3 *T.* 'Spring Green' (Viridiflora)
A classic green and ivory tulip, which is very reliably perennial. A really long-standing cottage garden plant.

4 *T.* 'Groenland' (Viridiflora)
This is the pink and green version of 'Spring Green' that we've grown at Perch Hill for five years or more and it still comes up in the same place.

5 *T.* 'Artist' (Viridiflora)
Orange flashed with green, this has slightly shorter stems than others in the Viridiflora group. It may be the most perennial of the lot in our garden. It has come up around some roses for at least a decade.

6 *T.* 'Mistress Mystic' (Triumph)
The only Triumph here is this subtle grey-pink coloured tulip which has naturalised beneath our apple trees, forming decent-sized clumps (that produce more than one flower) in just three years.

7 *T.* 'Green Star' (Viridiflora × Lily-flowered)
A cross of the two greats result in a slim, elegant and architecturally shaped tulip in ivory with green strips.

Cut-flower tulips

All tulips make good cut flowers, but some are exceptional. I'm always on the lookout for varieties that have scent, out-of-the-ordinary glamour and excellent vase life.

As a general rule, it's the Doubles that have the longest vase life. They don't always have the prettiest proportions, but once cut and arranged, they are marvellous. *T.* 'Exotic Emperor' and 'Ballerina' are both here again! They have great shape and are beautiful en masse in a jug or as a single stem and – of course – 'Ballerina' is scented.

To maximise stem length, we occasionally pull our tulips with an abrupt yank straight upwards. The aim is to sever the stem from the bulb, often with some of the core of the bulb coming with it, which gives you a good long stem but kills the bulb. This is how cut-flower tulips are usually harvested in the Netherlands, but it seems pretty wasteful to me. Another option is to select tulips that are naturally tall, such as any of the French group, or the Lefeber hybrids, like 'Temple of Beauty'.

Tulips continue to grow once they're cut, with the main growth plate just below the flower. With heavy-headed Parrots, whose flowers tend to bend downwards after a day or two, puncture the top of the stem with a darning needle or thick pin – this disrupts the growth plate, slows cellular division and prevents the droopy head.

1 *Tulipa* 'Brown Sugar' (Triumph)
A marvellous tulip with a honey and freesia scent and a hint of garam masala.

2 *T.* 'Antraciet' (Double Late)
This satiny carmine has been one of my favourites for years. I want a dress like this.

3 *T.* 'Black Hero' (Double Late)
This is a rich crimson-black and properly glossy. I always think it should really be called 'Black Beauty'.

4 *T.* 'La Belle Époque' (Double Late)
An unusual, coffee-mousse coloured variety; an amazing glamour-puss in the garden and vase, and seems to mix with almost any colour.

5 *T.* 'Pink Star' (Double Late)
Like a peony, this is marvellous as a single stem, lovely in a cobalt blue glass vase. You wouldn't know it wasn't a peony, but yet it's there in April, flowering away: it fools everyone.

6 *T.* 'Rococo' (Parrot)

Several of the Parrots make even better cut flowers than garden bulbs as they are almost too glamorous and feel out of place. 'Rococo' (and 'Amazing Parrot') can do both and are favourites here.

7 *T.* 'Temple of Beauty'

A whopper to tower over you from a vase. Its height leads to some great twists of its stems.

Practical
April

There's so much going on in April, and so many jobs to do, that I often have to consciously carve out time to relax in the garden.

Snowdrops and aconites need to be lifted and divided (see page 197), and the hardy annual seedlings (such as *Salvia viridis* blue-flowered, *Ammi majus* and the annual scabious, *Scabiosa atropurpurea*), can be put out. We can do this even at the start of the month, as it's unlikely to get so cold in Sussex that they're at any risk. If the temperatures look mild, we try to harden them off quickly, moving them to a cold frame with the top open for a few days before planting them out.

Any undercover sowing that we didn't get around to in March is now urgent, with quick-growing half-hardy annuals tithonia and even cosmos still happy to be sown now. Around the middle of the month, we sow zinnias in guttering. Before then, it's just too cold for them.

Zinnias struggle if sown too early, but we always know when it's the moment to plant them, as it's the time we start to regularly eat supper outside. Zinnias hate cold nights and won't grow, but if we can sit comfortably in the garden in the evening, zinnias can too. So time your sowing to about four weeks before your usual first outside supper, (for us, that's now).

You can direct sow now, too, but we do very little of that here. The germination is too sporadic on our heavy soil.

With all our dahlia tubers having been potted up for a month or so (see page 85 in March), they're now beginning to shoot well,

so it's time to thin them out, and if we want to propagate, now is also the time to take cuttings.

Weeds such as bittercress and groundsel can come up to flower fast at this time of year, so we keep our eyes peeled for them as we walk around – we hoe off (severing the tops from the roots of annual weeds), and mulch to help avoid any more weed germination as the soil warms.

All these practical jobs are important, but it's staking and creating support structures for plants and pots all round the garden that really absorbs our focus in April, and it's a job that we particularly love doing.

Staking and Support

Does this scenario sound familiar? After a night of wind and rain, you go out to the garden and find a beautiful delphinium lying flat on the ground, but you don't have time, right then and there, to stake it. Within a couple of days, the new growth – the axillary buds and the leading stem – have bent their tips towards the light at right angles. You then hoik the whole thing up and support it – slightly clumsily as it's now more difficult with the stems contorted – using a frame of canes and string. Within a couple of days, the growth tip turns up to the light again, not straightening the curve formed at the collapse, but creating another right angle. The plant is beginning to look more like a corkscrew than the lovely verdant thing it could have been. Oh what a shame!

While acknowledging that time can be hard to come by, as good gardeners, we want to avoid this. You can use a quick and simple cane and string support (see page 194 in June), or go to town and create structures that are both beautiful and architectural. Like garden statues or collections of pots, they add an extra dimension to the garden. We're rather obsessed with this at Perch Hill and have a healthy level of competition amongst the team, all striving to erect the best, the biggest, the most architectural and, of course, the most functionally effective frames.

Frames for borders

Staking tall-growing plants is key to a beautiful garden, and April is just the time to tackle a perennial border. At this moment in the year, herbaceous perennials have emerged from the ground but are still compact and low and haven't begun to flop about. You can see the whole crown, making it much easier to enclose within the frame. By May, the plants may be tall enough to collapse on their sides and you may have missed your moment.

All the early flowerers such as peonies, hardy geraniums, oriental poppies, lupins and delphiniums will look magnificent in summer if you stake them now, and while you're at it, stake later-flowers such as heleniums, phlox and cardoons. For heavy-stemmed plants with large flowers, such as dahlias, the taller varieties of sunflowers, nicotiana and chrysanthemums, it's best to stake each plant individually (see page 194).

If you can get your hands on some hazel or silver birch pea sticks, or some coppiced willow, you can create some beautiful frames. Unlike many of the bright, shiny bamboo canes, these twigs and branches are all renewable and local, not flown in from thousands of miles away. The Sussex countryside is full of hazel and silver birch and coppicing it is having a renaissance.

Willow is tricky on our heavy soil, as it tends to root, but you can decrease the chances of this by debarking the stake to 15cm (6in) above the ground. When these branches are woven into structures both tall and short, they do a good supporting job and look wonderful in the process.

The frames give the young, spring garden good sculptural presence until things really start to grow – and making these frames is easier than it looks. We make most of ours from birch, but also use hazel tops – with almost flat, fan-like rays of twigs, they make brilliant staking material. You can poke them firmly into the ground to enclose smaller things, such as *Alchemilla* and perennial geraniums, and bend a few of the twigs in toward the centre to make a nest, which the new growth can come up through. You can also use taller hazel ends, standing around 1m (3ft) high, to support stately grasses such as *Miscanthus*; again, create a web-like nest from the twig tips over the centre of the clump.

If you don't have access to hazel, use the prunings from twiggy shrubs from your own garden. Mock orange is excellent and weigela is pretty good for smaller plants, even the not-so-twiggy bits can be used to make an inward-pointing lattice surrounding something floppy. That said, don't prune your weigela or mock orange in the spring or you'll be cutting off all your flowers. Below are instructions for creating a dome, and there is a step-by-step to creating a teepee (ideal for sweet peas) on page 86.

Creating a dome structure

❧ The height of the branches you select will depend on the height of the dome you want to make. As a rule, the branches need to stand about twice as tall as the final dome. And the dome should be about half the height of the final height of the plant you're staking.

❧ At 30cm (1ft) intervals, stick an upright well into the ground around the plant. We use this system, or similar, for many of our euphorbias, phlox, heleniums and delphiniums, as well as for blocks of annuals such as cosmos.

❧ Curve the first branch almost at a right angle at the height of the top of the dome. If harvested when the sap starts to rise, early in the New Year, silver birch will do this easily without snapping – it is surprisingly pliable and you can treat it almost like rope or string.

❧ Carry on bending each branch over, binding it to the branch opposite until you've created a dome.

❧ To finish, gather the twiggy side branches and create horizontal rings binding not across, but around, in ideally a double layer to give strength and extra support.

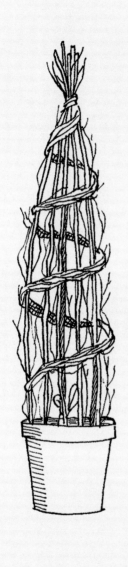

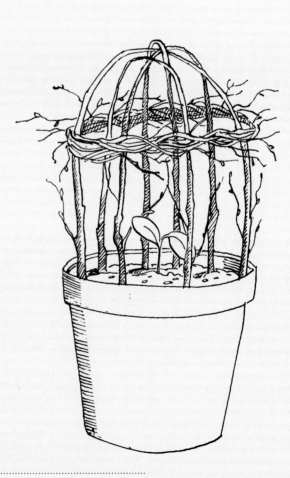

Frames for pots

For temporary height to line a path and transform it into an avenue, you can create teepees and domes to place in pots, which instantly increases their scale. The teepees and domes are created in just the same way as those in the ground. Plants ideal for growing on these raised climbing frames are half-hardy annual climbers such as purple bell vine, black-eyed Susan and Spanish flag (see page 244).

Pot teepee

* If you're making a teepee for a large pot, use 6 to 8 uprights about 2m (6ft) tall and push them very firmly inside the edge of the container. You'll probably need to work from a ladder.
* Gather the uprights together with a flexi-tie or twine at the top.
* Gather all the side twigs of a branch into a bundle in your hands and twist and bind them into a band, moving round from one upright to the next. You can do this in 2 or 3 horizontal layers or in a continual swirl.
* If you use canes, or your branches aren't twiggy enough to make a real witch's broom, add smaller sticks between the uprights, or attach string at the base of the uprights, to give the plants enough handholds to climb.

Pot dome

* To make a dome, arrange 6 to 8 uprights standing vertically inside the edge of the pot.
* Choose a pair of branches that are opposite to each other and bend them down, then twist them to secure them onto each other. Continue to pair up the uprights, bending them one below and one above the previous pair, until you have bent all the verticals into a dome.
* Finally twist horizontals of willow or hazel 15cm (6in) above soil level to encircle the whole frame, twisting in and out with several lengths until the frame is firmly bound together.

Dahlia Cuttings

Dahlia cuttings are a great way to increase your dahlia stock. Dahlia tubers will produce many shoots in spring, and you don't need any more than five to seven shoots per tuber to give you an abundantly flowering plant, so if you get more shoots, use them for cuttings.

These will be established enough to plant out in the garden about six to eight weeks from striking. They will then grow so rapidly they're almost indistinguishable from the mother plant by the end of their first season.

They will form decent-sized tubers that can be mulched and left in the garden, or lifted, as you do with mature dahlias, depending on the climate.

These first-season rooted cuttings will flower a little later than the mother plants and will continue on a little longer too. Flower farmers use this method to extend their dahlia season.

Taking dahlia cuttings, which are then pushed in around the edge of a pot.

- Slice off short, stout shoots with a sharp knife, aiming to slice a little of the tuber with the cutting if you can.
- Leave just the topmost leaves (below the tip) and remove any leaves down the stem.
- If this top pair of leaves is very large, trim them to half the size to reduce the need for water uptake (for photosynthesis and transpiration).
- Pinch out the tip. By doing this, you remove apical dominance and the growth hormone is diverted from pushing the plant upwards, and instead channels into strong root growth.
- Push the cuttings into a gritty mix of compost (1 part grit to 3 parts compost), spaced 2½cm (1in) apart, around the edge of a pot. Water and label. They will root faster with a bit of basal heat, so put them on a propagator mat (see page 79) if you have one.
- They will have rooted well within a month. At this point, knock the pot of cuttings out onto a bench and repot each cutting individually into its own 9cm (4in) pot.
- Put them back on the heated mat for another two weeks. Then repot into a pot one size up.
- Plant out once the chance of frost has passed in late May or early June.

Chrysanthemum Cuttings

In general, you should follow the same guidelines as for dahlias, but there are a few tips to bear in mind.

❧ Take your cutting, ideally 5-8cm (2-3in), from the base of the plant, trying to get a little of the mother plant's root if you can.

❧ As with dahlias, insert these around the edge of a pot, and once rooted, pot them on.

❧ To get our chrysanthemums looking their best, we pot on three cuttings of the same variety to a large pot (around 7-litre) and sink the whole pot in the ground. There it stays until autumn when, ideally, it gets lifted and brought inside to extend the flower-picking season.

❧ As with almost all plants you grow for picking, you want stocky, stout front-row forwards rather than spindly wingers. This gives you the best production later in the year, so a couple of weeks after potting, pinch them out, removing the top growth down to three or four leaves (about 22cm/9in tall).

Direct Sowing Hardy Annuals

We do very little direct sowing at Perch Hill as it really does not work very well on our heavy clay soils. However, there are a few exceptions, particularly plants that struggle with root disturbance. Dill, nigella and poppies (all except Iceland poppies; see page 192) are direct sown in April.

❧ Direct sowing works best on free-draining soil and suits plants that germinate quickly and easily once the soil is warm.

❧ You want the soil to be warm and moist, as it is in April and September, and have a fine tilth. Don't worry too much about them: hardy annuals are able to survive outside and grow even if there is frost at night (unlike half-hardy annuals).

❧ Sow as thinly as you can, individually placing seeds if possible.

❧ Wait until the seedlings are 2-3cm (1-2in) tall and thin them (use the spacing guidelines on the packet of each plant). Leave them to grow on until they flower and then pick, pick, pick.

❧ For any varieties that grow to more than 45cm (17in), stake them with jute netting or silver birch once they get to about 30cm (12in), and they will quickly grow through the netting and be supported. Their leaves will cover the netting and it will disappear.

May

This patch of East Sussex is thick with woods, and at this time of year the leaves of the trees are brand new and perfect, the best being the sharp, yellow-green of the oaks, which completely encircle the garden. I remember someone saying to me, May is like a party where the dress code is green. And that's certainly true where we live. The green of the oaks dominates, together with the bronze-green hawthorn hedges, the matt-green bluebell leaves and the almost fluorescent hue of the new grass along the drive and in the fields.

The verges along the lanes and the edges of the fields all around us are full of cow parsley and hawthorn blossom in May, and the woods are carpeted with wild garlic that merges with bluebells on the dryer ground. I love to pick an armful of cow parsley. I bring it home from a walk and sear the stem ends in boiling water for 20 seconds and then leave the bunch in cold water for a couple of hours before arranging – this transforms cow parsley into a cut flower that will last almost a week (see page 230 for cut-flower conditioning). Just make sure you don't pick hemlock, which has vertical purple splotches on its stem and is very poisonous.

The lane we live on looks like the best natural garden and trumps pretty much anything we gardeners can do, but with that in mind, we have made a wildflower meadow right outside our kitchen. We have a big window facing west towards the meadow, and beyond that, a great view. There's a shepherd's hut there, so when the nights warm up, we sleep out in the middle of the meadow, at its best in May.

Looking from the Dutch Yard across the Oast Garden in mid-May, with *Allium hollandicum* and *A. hollandicum* 'Purple Sensation' drifting all through, and *Elaeagnus* 'Quicksilver' glowing beautifully in the foreground.

In the garden proper, many of the tulips are starting to fade, but we try to make the most of those that remain. The huge-headed Parrots are looking brilliant and I love to arrange as big a vase of them as I dare. The tall, curvy, purple Parrot tulip 'Victoria's Secret' contrasted with 'Amazing Parrot' is as dazzling as you can hope for. 'Flaming Parrot', growing on its own in a pot or cut for inside, is equally marvellous.

In the Oast Garden, alliums are opening to replace tulips, and with them, carpets of self-sown aquilegia. We have lots of double hybrids in a range of purples and crimsons, originating from the parents, 'Ruby Port' and 'Blue Barlow', planted maybe ten years ago, their descendants all now mixed into a range of colours, with some tall single varieties, like the wild woodland common columbine (*Aquilegia vulgaris*), in there too.

All my favourite biennials are coming into their own in May. Iceland poppies, foxgloves, sweet Williams and scented stocks join the wallflowers and honesty, which have been flowering for a month or so.

Below An armful of foxgloves in late May.
Opposite Foxglove 'Sutton's Apricot' underplanted with *Allium hollandicum* 'Purple Sensation' and *A. cristophii*.

The Perennial Cutting Garden in late May, with sweet William 'Electron Mix', emerging delphiniums and *Stipa gigantea*.

Autumn-sown hardy annuals – honeywort, marigolds and euphorbias – are just as brilliant and I fill our vases with lots of those, every stem seared in boiling water first. There's also the odd early peony just starting to flower.

It's light now into the evening and with the frosts over, it's peak planting time. Pretty much every moment of every day we are busy planting the seedlings and cuttings we've grown and rooted. It's also time to plant out the dahlia tubers (see page 85), which have been growing away. This is a time of major transition, with all these things steadily filling the garden for its summer and autumn jamboree.

Opposite *Smyrnium perfoliatum* and *Allium hollandicum* 'Purple Sensation'.
Below *Geum* 'Mrs J. Bradshaw' with self-sown alliums and aquilegias in the Oast Garden.

Wildflower meadow

Wildflowers are the reason I've spent my life gardening, so it's no surprise that when we arrived at Perch Hill, creating a wildflower meadow was one of my priorities.

I'd been amazed by the meadow in Christopher Lloyd's garden at Great Dixter, where I'd briefly worked as a volunteer. The grass is dusted with flowers, starting with crocuses in February, mainly *Crocus tommasinianus* then golden crocus (*C. chrysanthus*), and also snowdrops around the shady edges. After the crocuses come narcissi and enviable quantities of snake's head fritillary almost to the end of spring. Scattered through, there's also a rich procession of orchids, starting with early purple, then common spotted, then the rarer twayblade and green-winged orchid, which all happily self-sow, as well as the odd bee orchid, tucked in.

Achieving this level of succession throughout spring and summer until a late grass cut in July or August, has taken more than a hundred years at Great Dixter. Bulb planting there started in the early 20th century with Christopher Lloyd's parents, and self-sowing and a lot of dividing of bulbs has been going on ever since.

I knew we couldn't jump straight to having a thriving wildflower meadow, but I was determined to make my dream of at least some flowery grass come true. It's turned out to be much more difficult than I thought, and while we couldn't recreate Great Dixter, we do have a flowery meadow of sorts.

It was something of an uphill battle because of the rich soil here. I couldn't face stripping off the topsoil to get down to pure clay (which is what some people recommend), so I started by sowing tons of yellow rattle. This is an annual and a hemiparasite, which means that it fixes its roots onto the root system of an adjacent grass to extract the water and minerals it needs. This weakens the grass (the main competition) and allows more delicate and often more colourful and interesting wildflowers to do well.

In his book, *Where Have all the Flowers Gone?*, meadow restoration expert Charles Flower writes about a small meadow on his farm in Wiltshire. This area used to produce 240 bales

Our wildflower meadow with *Narcissus* 'Hawera', red campion and ragged robin. This is the view from our kitchen window, looking west to the far side of the valley.

of hay. After introducing yellow rattle and allowing it to fully establish for a few years, his count of bales dropped to 90. You can see why farmers would have considered yellow rattle a pest, robbing them of precious grass to feed their animals in the winter. But for those of us keen on wildflowers, it is just the thing.

Sadly, on our coarse grassland – mainly made up of cock's foot, perennial ryegrass, meadow foxtail and Yorkshire fog, with very little red fescue, the rattle's main host grass – even this vigorous annual has struggled and we didn't get the transformative effect I had hoped for. I have tried with yellow rattle in four or five places now, but it's yet to really take hold in any of them, the strong grasses just romping back when I stop hand-sowing the rattle.

This failure to thrive may also be down to the fertiliser and herbicide residue here; Perch Hill is known historically as the poorest farm in the parish. To make up for that, before our time, it was subject to a generation of chemical additives. Chemicals are very persistent on poorly drained clay soils. They stick around, rather than being washed out into the water table and local rivers.

As an alternative, in our important and prominent areas (such as outside the kitchen), we resorted to wildflower turf – a bit of a

cheat's solution that gives faster results and one I'd recommend for rich soils. I selected wildflowers within the grasses (a mix of fescues and crested dog's-tail) that were robust and prolific enough self-seeders to hold their own in the turf.

The turf established really quickly, in fact pretty much straight away, with flowers almost instantly, and each spring the mix is dominated by campions. We have a smattering of white campion, lots of red campion and also ragged robin doing well. We have introduced cowslips in plugs and they are holding their own, but not self-seeding much. The campions flower throughout May, with ox-eye daises and sorrel taking up the baton in early summer. We have to keep an eye out for creeping buttercup, which would carpet the whole place given half a chance (it needs to be dug out if more than a few plants appear). Then in July, and until we cut it, the grass is dominated by wild carrot, alongside hoary and ribwort plantain.

These wildflowers are hugely important for the view from our kitchen in spring and summer, and we've extended the seasonal show with bulbs studded through. For February, we have *Crocus tommasinianus*, which quickly self-sows, and a succession of narcissi for March and April. There are lots of *Narcissus* 'Actaea',

Wild carrot froths up in the wildflower meadow in July.

as well as *N.* 'Silver Chimes'. In May we have the classic *N. poeticus* var. *recurvus*, and also *N.* 'Hawera', both late-flowerers, highly scented and beautiful.

From the middle of March, inspired by Cretan bulb fields (see page 96 in April), we also have a succession of wild-looking tulips. The earliest pair is *Tulipa praestans* 'Shogun' in a golden yellow, and *T. turkestanica*. Then we have *Tulipa* 'Peppermintstick', which coincides with *N.* 'Xit', a new variety that has become a favourite. We have planted snake's head fritillary, but it's difficult to tell whether it's naturalising as the flowers are still sparse.

In May, our campions give the flowers along the lanes around East Sussex a run for their money, and there's the added joy of camassia clumping up and thriving on our heavy soil. We have the standard blue *C. leichtlinii*, as well as the lovely cream, 'Alba'.

The wildflowers really take over at this time of year, and you wouldn't see the most delicate bulbs. But after we've cut the grass in late July, we get *Crocus speciosus*. This is planted in autumn and flowers the following September when the grass is short, so the purple starry flowers are in clear view. We tried planting the autumn-flowering colchicums, but now avoid them – our wildflower meadow is in the most exposed place at Perch Hill and the wind reduces their flowers to tatters within hours.

There are a few tricks for getting all of these bulb varieties looking really good in grass. The first is to plant them in swathes, not patches. It's all about numbers. With each variety, it's best to wait until you have the money and energy to plant as many as your space can handle. In a sizable area, this can mean hundreds of each variety, or if you have a city garden, thirty of each. The second trick, if you want to aspire to Great Dixter, is to go for a couple of different varieties each year and plant them en masse, then go for something else the following season. Finally, when you're planting, a random distribution is key. We follow the method of moving through four metre squares of lawn, then standing in the centre of each, chucking handfuls of bulbs up in the air to be planted where they land (see page 335 in October).

Little-by-little, the wildflower meadow, so prominent in our day-to-day from the windows, gets prettier for longer. It's not Great Dixter, but we do have a loose lawn that's pretty full of flowers for at least seven months of the year.

Narcissi for grass

Since trialling the best narcissi for cutting in the 1990s (see page 66 in March), we have trialled narcissi for grass, growing them in three different places to assess how they perform. We have some down the drive, some in our wildflower meadow and more in the damson orchard.

Particularly on heavy clay, narcissi (along with camassias) are by far the easiest, most perennial and successful bulbs for grass. All the ones in the list here have lasted in the garden for years already and will go on for decades. That's the brilliance of narcissi when compared to tulips: they are true perennials. At our house in Scotland, we have narcissi that were planted by the Victorians that still flower away happily. Clumps of all the plants listed on the following page have come back bigger and better each year, so much so, we have now lifted and divided them several times (we do this after flowering, when they're still in the green; see page 197 in June).

To score highly for us, bulbs in grass need certain characteristics. First, they must look natural, as you might see them in the wild. (We were sent some doubles by mistake once and planted them in the lawn: we took them out as soon as they flowered, they looked all wrong.) Second, are fine leaves; these are a must-have. Coarse leaves don't look good in grass as they die back, so chive-like foliage is the goal. Finally they need to be perennial and good at naturalising, but with narcissi that's pretty much a given. Scent is a bonus, so we can pick them as cut-flowers too.

Narcissus 'Actaea', flowering from April into May in shade on the curve of our drive.

Best of the narcissi for grass

Listed in order of flowering from March until May

Narcissi stud all our areas of rough grass in March and April, but it's in May when my two favourite varieties come into flower, Narcissus 'Hawera' and N. poeticus var. recurvus – that's why I've included narcissi for grass here, rather than earlier in spring.

1 *Narcissus pseudonarcissus*
The wild British native daffodil that will spread by self-sowing beneath trees, shrubs and in grass too. It's one of the longest lasting, easiest and most successful for naturalising.

2 *N.* 'Trevithian' (Jonquil)
'Tête-à-tête' is an obvious choice, but I don't, on the whole, choose yellow daffodils, particularly for planting in grass. This golden-yellow 'Trevithian' is the exception. It has a jasmine-like scent and very fine leaves.

3 *N.* 'W. P. Milner' (Trumpet)
Perfect for a position right at the front of the border, this dwarf Trumpet has handsome, washed-silver foliage. Lovely in the garden and in containers, and a favourite for growing in grass where it will gradually spread into delicate-looking but robust and long-flowering carpets. It is not scented.

4 *N.* 'Silver Chimes' (Tazetta)
Delicate leaves, pretty flowers and nice scent. This looks good and does well in grass, bulking up fast to form good-sized clumps.

5 *N.* 'Xit' (Small-cupped)
A new discovery and absolutely beautiful: dainty, scented, fine leaves, long-flowering – it has it all.

6 *N.* 'Bell Song' (Jonquil)
Flowering from March into April, this has an unusual, peachy coloured corona that I love, and it has a pretty scent to match.

7 *N.* 'Hawera' (Triandrus)
Delicate, almost butterfly-like, petite, highly scented and a soft primrose yellow – this has naturalised well in our lawn.

8 *N. poeticus* var. *recurvus*
Planted in the wildflower meadow and down the drive, this has now done what I hoped and merged into a carpet of delicate, beautiful and highly scented flowers (pictured on page 149). *N.* 'Actaea' (pictured here and page 145), is also a poeticus variety – it flowers about a month earlier and is very similar, though a touch less delicate.

Narcissus poeticus var. *recurvus*
(known as old pheasant's eye)
with red campion in the wildflower
meadow in early May.

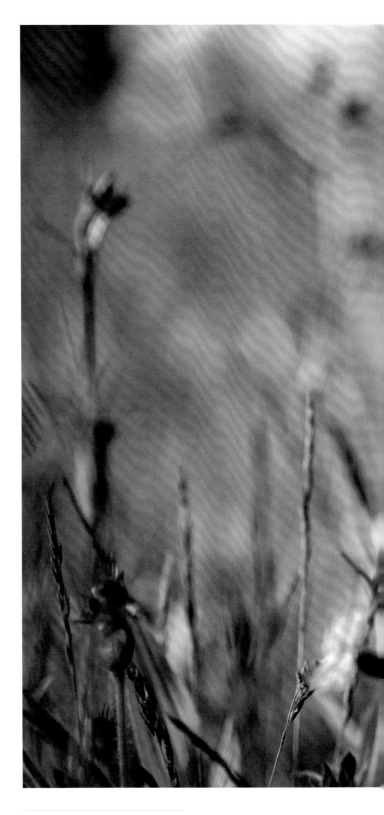

Alliums

Some accuse alliums of being storm-troopers, garden brutes which dominate at the expense of almost every other plant, but there's nothing like them for their sparkler explosions of colour – and they are invaluable in May. With tulips mainly over and roses and perennials yet to get going, we need a big player to give the garden some oomph through May. That's where the allium comes in.

To guarantee a May succession, we start off with the mauve, rather than saturated purple, *Allium jesdianum* 'Early Emperor', which opens early in the month when tulips are still going. *A. hollandicum* 'Purple Sensation' is the crucial variety for us, preventing a lull in full-on flower and colour. You can almost time it to the day when the last of the tulips (usually 'Victoria's Secret', 'Green Wave' and 'Amazing Parrot') drop their petals and a good band of 'Purple Sensation' turns properly purple.

I first planted fifty bulbs in the Oast Garden more than twenty years ago, adding half a trowel of grit to each of their planting holes. The bulbs thrived in this walled, sheltered spot and naturalised quickly. The danger is, over the years, the seedlings can build up to the exclusion of other plants. That happened here, so we regularly thin them out: every three or four years we dig up whole allium sections from the beds and get rid of them.

Allium leaves are one of the first to emerge in spring, with chives the classic example, poking up in February. By the time 'Purple Sensation' comes into flower, the basal leaf rosettes are turning yellow and don't look good. With the foliage starting to die back at the tips at almost the same moment the flowers fully open, it rather blights the triumphant effect, so we try to defoliate (remove the leaves) in mid-May. This early emergence of the foliage means they have had a good long go at photosynthesis before the demands of flowering, so leaf or no leaf, the bulb will have made enough food to thrive and grow into the following year, so total defoliation at this stage is fine. En masse, it's quite a job, but it's worth it.

Just so you know, allium leaves, like onion skins, stain yellow, so don't pick or defoliate alliums in your pristine whites. I've ruined a

In the Oast Garden picking alliums (*Allium hollandicum* 'Purple Sensation' and *A. hollandicum*, which is what 'Purple Sensation' reverts to when it self-seeds).

favourite linen dress doing just that. This all makes alliums sound like a hassle, but the Oast Garden would be a dull and colourless place in late May without them. They just need managing.

With the new generation of self-seeders comes a colour range from deep purple (of the planted hybrid, 'Purple Sensation'), to a softer mauve of the natural species *A. hollandicum*, to which the hybrid can revert. I didn't like that to start with: the Oast is meant to be about deep, saturated colours and there was a bit too much mauve for my liking. I've now relaxed and come to value the tonal variation.

All alliums are marvellous for pollinators. As a genus, they're one of the richest nectar sources – and that's important in May when hives are busy reproducing. The Oast Garden is positively manic with bumblebees and honey bees at this time of the year and I'm told by beekeepers that the oniony taint doesn't come through in the honey.

At Perch Hill, we fill the gaps between our allium spheres with plants such as the acid-green form of tansy, *Tanacetum vulgare* 'Isla Gold', as well as black-leaved cow parsley (*Anthriscus sylvestris* 'Ravenswing'). There are carpets of bronze fennel and my favourite of all spring foliage plants, the Cretan wild flower, *Smyrnium perfoliatum*, in the brightest green you'll ever see.

Verbascum 'Cotswold Beauty' with *Allium* 'Purple Rain', growing up through a frothy, smoky cloud of bronze fennel.

We also intersperse the allium drumsticks with saucers of poppies and marigolds, and spikes of foxgloves and mullein, but none of these have the om-pa-pa of alliums. They are the string section, rather than the trumpets of our May orchestra.

Opening about the same time in the farmhouse and in the Annual Cutting Garden, all the way down both sides of the main path, are groves of *Allium* 'Purple Rain'. This has a more exploded flower, airier in its form than 'Purple Sensation' and flowers for slightly longer. It has become another favourite.

Next comes *A. schubertii*. Its flowers are vast and a little shambolic, but also glamorous and alluring, and surprisingly for an onion, they smell sweet. The flower spikes erupt with the bud held in a tight, green mini-ball. Then overnight they explode, a halo of sparkler stars, one globe about 7½cm (3in) from the heart, and then another more distant constellation standing out on stems about 20cm (8in) long. Together, they create a sphere 40cm (16in) across when the stems finish growing. We've had a good clump in the same spot with us for twenty years. It's planted in a low section of the garden that has twice been flooded by a blocked ditch. The huge bulbs sat in water for more than a fortnight, but survived. They're truly reliable and don't self-sow, which some would say is a blessing.

A. schubertii has been crossed in recent years with another stalwart, *A. atropurpureum*, and the result is the deep purple *A.* 'Spider'. We've only had this in for a couple of years, so I cannot vouch for it in the same way as its two parents, but knowing them, it's likely to be just as reliable and glamorous, as well as very perennial.

The final favourite we have is the true drumstick *A. sphaerocephalon*. These are airy, the stems elegant and slim with a compact deep purple globe at the tip. This is another self-seeder and will even creep into cracks in stone or brick paths, so place this well back in the border if you like a tidy garden. I love their naturalistic feel.

With the larger-headed alliums, don't forget they are almost as good in seed as they are in flower. In July or August, we consciously go on a harvest and bring in as many allium heads (plus stems from which you can then hang them) as we can.

Once picked, we hang them upside down to dry. At Christmas, we spray them with paint, usually silver, which makes the brittle seedheads much less fragile. After Christmas, we store them in the attic (wrapped in tissue paper so they don't stick together) for the following year. *A. schubertii* looks marvellous at the top of the tree, and together with smaller varieties, adds a natural top-to-toe sparkle.

Best of the alliums

Listed in order of flowering

The ivory and pale pink alliums are increasingly fashionable, but it's the rich purples I still hold dear. They have so much more impact, and for us colour-lovers, they are not to be missed in May and June.

1 *Allium hollandicum* 'Purple Sensation'
The best all-round garden and flower-arranging allium. It starts to flower in mid-May. Will self-sow.

2 *A.* 'Purple Rain'
This comes into flower earlier than most (early May) and is still looking good in mid-June. The foliage is also quite fine, but even so, we tend to do some leaf removal in early May.

3 *A. stipitatum* 'Violet Beauty'
Sturdy and stocky, so often grown as a cut flower, this has medium-sized dense half-domes, ideal for vases and hand-tied bunches.

4 *A. cristophii*
I'm always surprised by how cheap these bulbs are given the massive effect they provide, with heads that get bigger and bigger over time. They've already been here for more than a decade with some flowers now bigger than a football.

5 *A.* 'Lucy Ball'
This variety gives great regularity and impressive stature, standing at hip height in drifts, giving borders massive oomph in late May and June. It's cheaper than *A. giganteum*, but just as perennial and both early and long-flowering.

6 *A. schubertii*
This crazily huge allium is soft mauve with seedheads that are as spectacular as the sparkler globes in full flower. It has been coming up in its original spot here for at least twenty years.

7 *A.* 'Spider'
This cross of *A. schubertii* with the famously perennial *A. atropurpureum* is a magnificent deep purple.

8 *A. sphaerocephalon*
Arriving one month after most other alliums is this small-headed variety. It's very pretty and will gently self-sow.

Practical
May

I seem to spend May with a trowel in my hand. We are very busy emptying the polytunnel of all the plants we've grown from tubers, seeds and cuttings. Apart from the odd pot or two, many people's gardens consist of mainly permanent planting, but that's not the case at Perch Hill. Practically every square metre gets a refit. There are permanent bones, but the colour for the next six months comes largely from tender plants that cannot go out into the garden until the frosts have passed in May. It's a lot of work alongside all of our trials – and we really feel it this month.

Planting out Half-hardy Annuals

All our half-hardy annuals need planting out as soon as possible. Bar our zinnias, the sooner we can get them out, the fewer pot-bound plants we'll have to deal with, so the pressure is on.

The first plants to go out are cosmos and autumn-sown antirrhinums. If we get a cold night or two, these would both survive, whereas if we'd planted zinnias out there, they'd be nailed. Towards the end of the month, they're the last thing to go in our wheelbarrow.

Pot-sown seedlings

In early May, we prepare our half-hardy planting spots. Increasingly we follow the no-dig system here, so add some organic matter on top of the soil if it's poor. We mix this with some grit in parts of the garden (such as the Annual Cutting Garden) that sit low and wet on heavy soil. The tender perennial climbers, such as cobaea and *Rhodochiton* should be planted in the same way. The main vines of these are often already well over a metre by now in the polytunnel and have outgrown their silver birch pot teepee frames (see page 88), so they're gunning to be planted in their final positions. Once the frosts are definitely finished for spring, we plant everything out. (For sowing in pots, see page 83 in March.)

❧ Fill a barrow or a water-tight trug with water, then plonk all your pots, standing upright, in there before planting. That means their roots will go into the ground good and moist, which will encourage them to settle in quickly.

❧ With a trowel, dig a decent-sized hole, aiming at twice the depth and width of the seedling's root ball.

❧ Water the hole so that the base roots will be in damp soil.

❧ Knock the seedling out of its pot with a firm strike to the pot base, tease out the roots a little if they're knotted, then plant.

❧ When planting climbers, tip them out of the pot, taking care not to disrupt the frame. Don't try to separate them, just lean the pot frame on to the garden frame and tie it in. This works well, with no root or top disturbance and uninterrupted growth. Already up to a good height, with everything left intact, they romp away.

❧ To keep the weeds at bay, cut down and keep soil temperatures more constant, we mulch

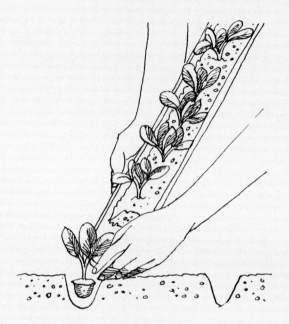

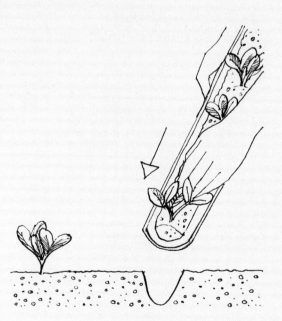

around our seedlings with 4cm (1½in) of green waste compost from our local council (or similar).

Gutter-sown seedlings

We planted half-hardy annuals such as zinnias in gutters in April. Once germinated and the seedlings are about 3cm (1in), they need planting out. To prevent the whole lot ending up on the floor, you ideally need two people to plant these, one at either end of the gutter. (For sowing in guttering, see page 82 in March.)

❧ Out in the garden, water the seedlings to glue the compost together.

❧ Slide the seedlings, plant by plant, straight from the guttering into the soil. Rather than doing this all at once, dig individual holes in the ground with a trowel, water the hole, then slide the first plant in. Then push the next plant forward to the mouth of the pipe, dig a hole for that, then slide that one into its hole and so on.

❧ Firm them in and water well.

Direct-sown seedlings

There are a few things – California, opium and corn poppies, as well as dill, tanacetum and nigella – that do better sown direct rather than under cover (see page 127 in April). Poppies in particular don't like being transplanted, but about a month after sowing, if sown thinly, they can be transplanted. If you do this carefully, lifting a really good ball of soil with each

Planting up a summer container of tender perennials, including pelargoniums, which are fine to go outside once the frosts are over.

Planting Up Pots

May is the moment we plant up almost all of our containers for summer and autumn, be it with plants grown from seeds or cuttings. It's change-over time, so aside from the first week of May, I have to take care not to teach my Magnificent Pots course at Perch Hill. Almost all of our containers are being replanted – and look far from magnificent. The tulip combinations come out and in go the tender perennial or half-hardy annuals, but they take a while to settle in.

We have quite a nifty system to avoid a total colour lull. We use robust plastic inner pots for our dolly tubs and long toms. The tulip lasagnes are planted in those (see page 365 in November/ December), as are the plants we plan to follow on with. So when the tulips start to look ropey, the whole inner pot is whipped out and the next combination is slotted straight in. Grown in the polytunnel, there are some things (such as nemesias, verbenas and argyranthemums) for the summer pots that we can force to be in flower as they go out in May.

The one downside of this succession system is the inner pot's black plastic rim. This can look ugly, at least until things have grown. Rather than propping them on the rim of the outer pot, we now allow them to sink down within it, by just an inch, supporting them from the base with empty, upturned plastic pots, empty compost bags or those awful polystyrene bedding strips we want to get rid of. This system is practical and works aesthetically, and – with half the volume

seedling, one row or block can easily be turned into three or four.

❋ With a trowel, try to dig up the seedlings with as much root and soil as you can. This minimises root disturbance at a critical time, which is key.

❋ Space them as recommended on the back of their packet – often, but not always, around 30cm (12in) apart.

❋ Water well in their new position.

of potting compost – is cheaper in the end too. We can't use this system with our largest pots for large climbers, or hungry feeders (such as dahlias), as we plant those straight into the pot to give their roots more room.

* Add crocks (or something similar) for drainage. Drainage is key, so all pots and containers need one or more holes in the bottom. The holes then need to be covered with crocks – pottery shards or pebbles – to keep them from clogging or blocking with emerging plant roots or compost.
* If you are planting up large pots, put crumpled empty compost plastic bags into the bottom. This provides good drainage and means you use less compost.
* For seasonal containers, we use a good quality, multi-purpose peat-free compost.
* Fill the pots with compost to within 2½cm (1in) of the rim and then create planting holes with a trowel or your hand. This is a better practice than placing plants into a half-filled pot and filling in with more compost around them. It's hard to get the compost into all the nooks and crannies around the plants' roots and you're often left with big air pockets.
* Pack the plants in at twice the density you'd space them in the garden. Then you get a better show and there's no room for weeds. That said, try not to overcrowd. Many of the seedlings or rooted cuttings will more than double in size by the end of the season.

* Water in well. Add labels to the pot edge so you can remember the exact names of the plants you've loved, or not, at the end of their season.

Planting Out Dahlias

Dahlias can go out in mid-May in the sheltered, warmer parts of the garden, and we wait until the last week of May to plant up new trial varieties in the more cold and windy parts of the garden. These can be planted out in much the same way as pot-sown seedlings (see page 157), but there are a few things to bear in mind.

* Dahlias thrive in a rich, moist soil, so it's worth adding plenty of organic matter to the planting position.
* With a spade, dig a decent-sized hole, aiming at twice the depth and width of the seedling's root ball.
* As you plant the dahlias, look out for any shoots that are starting to get leggy, and pinch them back. This promotes axillary bud formation and makes them bushy, giving you lots more flowers.
* Major slug and snail patrol is important when dahlias first go out. Dahlias are some of their favourite food, so concentrate whatever annihilation you choose around them. Treat nearby box, yew and garden hedges as much as you can. That's where slugs hang out in during the day, coming

Staking dahlias in a cage of willow
stakes and twine (see page 194
for how).

out to feast on your dahlias in the damp of the
night. Also be on frost alert and cover with
fleece if a late frost is forecast.

❁ Once dahlias have reached about 60cm (2ft),
and this can happen by late May, they need to
be staked. Make your own supports (see page
194), or use decorative iron frames. Simple iron
hoops will give enough support, but you need
one hoop for every plant.

June

In June, plants feel fresh, bold and vigorous. There's a perkiness to everything and a fecundity in the garden which is exciting, with flowers opening every day and the new things planted out in May filling the borders.

Every corner of the June garden and every June-gathered bunch can and should be fragrant. It's the month for roses, stocks and lilies, with teepees and arches of sweet peas coming into flower week by week. Every one of these plants pulsates with perfume. Wallflowers have been giving us a sweet scent too, which reminds me of the smell of macaroons. They begin to tire now, but can easily be replaced by the black sweet William, *Dianthus barbatus* 'Nigrescens', and for extraordinary vigour and a vase life of three weeks, we grow *D. barbatus* 'Electron Mix'.

Both wallflowers and sweet Williams are biennials, a group of plants we grow lots of here. Along with alliums, they plug that mid-May to mid-June colour gap. After the tulips and before the main border perennials really get going (delphiniums, agapanthus, phlox and summer-flowering clematis), biennials help to mitigate the lull.

Biennials are also useful for creating the essential rivers of connection between one large plant and the next, the all-important mortar to cement the different colour blocks into a cohesive whole. Easy to grow from seed and cheap as plug plants in tens, or even fifties if you have the space, biennials are perfect for creating broad brushstrokes of colour.

We have a soft, pastel colour theme in our Farmhouse Garden and use lots of the biennial creamy white foxglove (*Digitalis purpurea* f. *albiflora*) for just this filler role, as well as trays of its brother,

Opposite top The Perennial Cutting Garden in June with hardy annuals and biennials at their peak: sweet William 'Electron Mix', opium poppies and euphorbia, with ammi on its way.
Opposite bottom left What we refer to as the 'Perch Hill poppy', self-seeding ever since we arrived.
Opposite bottom right An Annual Cutting Garden June harvest of hardy annuals and biennials, sweet Williams, sweet peas, foxgloves, Iceland poppies, ammi, nigella and orlaya.

the soft, dusky foxglove 'Sutton's Apricot'. Its vertical spires are handsomely irregular, each with a slight twist and a relaxed ease. The pale pink is a tricky tone with many of the rich and bright flowers we use so much of here, but with the softer colours in this area of the garden, it's ideal.

Then there are our Iceland poppies, originally grown here as a cut flower, but now threaded all over the garden because they're so good. *Papaver nudicaule* 'Champagne Bubbles' look like spinning plates, with saucers standing high on twisting stems. Surprisingly, they're scented and attract large numbers of honey bees and bumblebees. Poppies of all kinds have abundant, protein-rich pollen, a key nutrient for a healthy hive, so we encourage poppies to make themselves at home wherever they can.

The dawn chorus may have quietened down a little by June, but everywhere you look there's a new group of blue tits, or a family of goldfinches, the young increasingly indistinguishable from their parents now. The number of garden birds at Perch Hill has multiplied vastly since we started feeding them properly. So much so, that parts of the garden can now feel like a natural aviary. And with the bird numbers increasing, the slugs and snails decrease. I now consciously listen for them, and rather brutally, love the sound of a thrush destroying a snail shell on the brick and stone paths.

A really precious thing is when the sun breaks over the still dewy garden, very early in the morning. I'm an early riser and this is my favourite time. And by June, the evenings are warm enough to sit outside regularly to eat, with a brazier lit for when the sun drops. And then, if we're lucky, we will hear a nightingale singing in the wood. They had disappeared from Perch Hill, but in the last two years, in one area in particular left to return to the wild, we now seem to have three or four males every year. My husband Adam and I also heard turtle doves recently for the first time, and combined with the call of the cuckoo, it makes a rare but uplifting trio of early summer birds to listen out for.

The Rose and Herb Garden in its first summer, with *Lychnis coronaria*, *Rosa* 'De Resht', rosemary, myrtle and olives all settling in.

Roses

My story with roses starts with two voluptuous pink beauties I came to know in the garden at Sissinghurst. The first is 'Madame Grégoire Staechelin', an early-flowering, loose, soft pink, scented climbing rose. The second is 'Madame Isaac Péreire', a Climbing Bourbon with full flowers, all ruffled and ruched, in a rich pink with massive scent. They are both the sort of rose a child would paint, a picture-perfect kind of rose.

As soon as I started to make my own garden, I planted both immediately. They are exceptionally lovely and I longed for them to flower, but the first Madame did so for only three weeks in late-May, and the second was so plagued by blackspot and mildew it defoliated entirely in its first three years and looked truly terrible, with only a smattering of delicious flowers.

The whole experience put me off roses. It seemed to me there were plenty of other plants which performed for ten times as long and didn't require the chemical regime we appeared to need to keep these grande dames happy and healthy.

A few roses crept into the garden over the following decade. I planted several of the ever-flowering China rose, *Rosa* × *odorata* 'Mutabilis', which I remembered from my parents' garden and rightly has the reputation of being super-healthy, if not super-showy. 'Mutabilis' also goes well with red-brick paths and walls, which we have in abundance at Perch Hill and is exceptionally long-flowering.

We also had to have 'Roseraie de l'Hay'. This was another rose I was brought up with; it has a strong scent and is the benchmark against which many people judge all other rose perfumes – so this crept in as a hedge and windbreak.

Adam made a plea for the carmine velvet 'Charles de Mills', and even richer, 'Tuscany Superb', which he remembered from his childhood. A garden designer friend, Pip Morrison, recommended 'Cerise Bouquet' – a huge, healthy, quick-growing Shrub rose in the most brilliant pink – and it's planted on the east wall of our bold and brilliant-coloured Oast Garden.

A bunch of rose 'Aphrodite', which we call 'the marshmallow'.

All five are still going strong, but that's as far as I went with roses, consciously avoiding being seduced. Visitors would often comment on the garden's dearth of roses, but as an averse-to-chemicals gardener, I remained wary.

Then Josie Lewis, our head gardener for the last seven years, arrived with a devotion to and knowledge of roses. With her, the rose count at Perch Hill has completely changed. From very few, we now have more than sixty varieties, ranging from climbers to hedge roses, huge species shrubs mainly grown for hips to lots of heavily perfumed bush roses in what we now call our rose garden.

When we restored our 19th-century hay barn, putting back its old Sussex peg-tile roof and strengthening its cracked and partially collapsing walls, we also decided to create a garden where our cold frames had once been on the south side of the barn. This is a 18m x 30m (60 x 100ft) space, sheltered on all sides, which wasn't being used to its full potential in our otherwise windy site, so we made a plan to fill it with herbs and roses.

Salvias 'Jezebel' and 'Nachtvlinder' underplanting and interplanting our roses ('De Resht', 'Tuscany Superb' and 'Felicia').

Josie steered me toward selecting varieties she knew to be naturally healthy and easy to grow, and we deliberately chose forms with not three, but more like twenty weeks of flowering. Most of

our roses flower from early June well into September (some having a break for a bit before restarting). Amazingly (to me) we often have roses still in flower in November, or even January, when they're pruned (see page 49).

So back in 2015, with a plethora of bare root roses about to arrive, we decided to do an experiment. There is an urban, or perhaps better described as rural gardeners' legend that suggests underplanting roses with alliums or salvias keeps fungal diseases at bay. Alliums are tricky: they are spectacular in flower, but their foliage quickly canopies over more delicate perennials and drowns them. We wanted to bridge the gap between tulips and roses with early-flowering perennials that would not have survived the allium foliage canopy – such as rock rose (*Helianthemum*), false indigo (*Baptisia*) and thymes – so alliums were banned from the rose garden.

Having visited the Dyson's Nurseries large salvia collection, Josie and I decided to go for a range of these to carpet beneath the roses. We selected salvias that are hardy (here in East Sussex, which can reach -10°C/14°F), relatively compact, shrubby types with pungent, aromatic leaves. This included *Salvia microphylla*, as well as *S. greggii* (including the Mirage series) and *S. × jamensis* hybrids.

Gardeners famously had perfect roses until the Clean Air Act in 1956 and '68, when levels of sulphur dioxide were hugely reduced, and with that, the level of fungal diseases on prone plants, such as roses and hollyhocks (also blighted by rust), hugely increased. I have a theory that salvias have good concentrations of sulphur in their scent profile (squash the leaves and you'll see what I mean), and that in the heat of the day they release some sort of natural (rather than toxic) fungicide into the air to keep the roses disease-free.

The salvias' hooded flowers are also stacked full of nectar, continually abuzz with pollinators. That's important to all of us at Perch Hill as they attract plenty of beneficial insects such as lacewings and ladybirds, whose larvae then feast on aphids. This also helps keep our roses neat and healthy.

Well, it's hats off to salvias, as all but the rose 'Munstead Wood' are still looking pretty healthy, even late in the year, without a sniff of fungicide. So on the back of this organic-gardening, companion-planting triumph, we added a 'rose for picking' garden. Of course, I'd always wanted to be able to wander outside and harvest enough roses to fill a bowl for our kitchen table, or a few highly scented stems to go in our bedroom. I had plenty of cut flowers to harvest

Powerfully scented rose 'Belle Epoque' with honeysuckle, lemon verbena and pelargonium 'Attar of Roses'.

and I could live without cut roses, but their addition in 2019 was a wonderful thing.

When it comes to roses for picking, we don't plant them singly, but try to make room for at least three of each variety, and we include pretty much the whole colour range. I didn't know I wanted a rose the colour of milky coffee (the sensational 'Julia's Rose' or 'Koko Loco'), or one in mustard that fades into grey ('Honey Dijon'), but these rose colours are now amongst my favourites.

It's worth picking roses early in the morning (though never in rain) and then – crucially – bringing them indoors to sear the stem ends in boiling water for 20 seconds. This extends the vase life more effectively than cutting at a steep angle. You can strip the thorns, but I often don't bother, though growing a few thornless varieties is handy. After searing, plunge them up to their necks in a bucket of cold water, ideally overnight, as this stops the stem end continuing to 'cook' and better adjusts them to the temperate change inside. When they're ready to arrange, add a mini slosh of clear vinegar or a pinch of bicarbonate of soda into the vase water to make them last even longer. (For more on conditioning cut flowers, see page 230.)

In the selection of my favourite roses the number one trait is, of course, a show-stopping, unqualified beauty. You want to be wooed by either form, colour or scent, and preferably all three.

Disease resistance is also key, so I can grow our roses well organically; I'd never go back on that. Our pollinators are threatened enough by agriculture without us gardeners reaching for a fungicide or insecticide spray.

Because I now love to cut roses, I'm also looking for reasonably quick-growing varieties with decently prolific flowering over a long season. We do grow *Rosa glauca*, *R. moyesii* 'Geranium' and the exquisite black-hipped *R. spinosissima* 'Merthyr Mawr', as well as *R.* 'Hansa', all four for their beautiful, healthy leaves and their impressive hips more than their flowers. And we still grow 'Charles de Mills' and 'Tuscany Superb', as I'd miss them for their June magnificence. But for the most part, a long and bountiful flowering season is the aim. The number of flowers per bush in one year produced by the terracotta 'Pumpkin Patch' and pure white 'Starlight Symphony' is staggering.

Best of the roses

The list is divided into three colour palettes: 'rich and brilliant', 'warm' and 'cool' – this latter group includes white and cool pinks (rather than the hot ones). Each choice within each group mixes well together, but you need to take more care mixing across the groups. Apart from in the roses-for-picking trial garden (where I rather love a crazy colour hullabaloo), we tend to stick to just one of these palettes in one area.

Rich and brilliant

1 *Rosa* 'Cerise Bouquet' (Shrub) One of the very best roses in my garden that flowers from May until September. No scent, but prolific, disease-resistant and long-flowering. I end up picking huge bunches, Constance Spry-style, and they have a vase life of three days. They are super-thorny, so cut with thorn-proof gloves.

2 *R.* 'Charles de Mills' (Gallica) One of the few varieties we grow at Perch Hill that flowers only once, but the velvet texture, along with the incredible scent, make this a must-have. It's healthy too, with elegant, bright-green, chiselled foliage that makes it a handsome looking plant, even without its flowers. It cuts pretty well, lasting about

three days in a vase. Vita Sackville-West wrote about how the runners are often even healthier and stronger than the mother plant and we have found this to be the case. Given how successfully it self-propagates (literally making new bushes a metre or more away), our plants are now just formed by the runners, and we have discarded the original trio planted more than a decade ago.

3 *R.* 'Cinco de Mayo' (Floribunda) An amazing red-brown colour, this is exceptionally long-flowering (May to December) with good scent, and is excellent for picking with a vase life of four days. It has big, multi-flowered heads on tall stems. Overall, pretty outstanding.

4 *R.* 'De Resht' (Shrub) An excellent rose for a small garden, which like 'Munstead Wood', is compact. It's also healthy and long-flowering lightly right into the autumn.

5 *R.* 'Hot Chocolate' (Floribunda) You'll either love or hate this. It's a favourite of both mine and Josie. It's the strangest copper brown, with a vase life of four days and a decent scent, and it's still flowering here in December. The downside is it's very thorny.

6 *R. moyesii* (Shrub) I wouldn't be without *R. moyesii* 'Geranium' (or 'Highdownensis', which has pink flowers) for their hips. These are like mini lightbulbs made of sealing wax, shining out through late summer and holding right on into the autumn. This is a key

12

plant in the Oast Garden, and its presence is crucial to the jungle-like, colourful look of this garden.

7 *R.* 'Munstead Wood' (Shrub)
I wouldn't be without 'Munstead Wood' – it's compact and beautiful for picking, a dark and velvety crimson, and unlike 'Charles de Mills' and 'Tuscany Superb', is repeat-flowering. The similarly coloured 'Darcey Bussell' slightly outshines it in health (Munstead succumbs to a bit of blackspot here, even with salvia underplanted), but we still love it for its general deliciousness, including its scent.

8 *R.* 'Rhapsody in Blue' (Shrub)
I love this for its cardinal purple colouring when it freshly opens, but maybe better still are the flowers as they fade, becoming a sort of washed-out purple linen. Big, multi-flowered heads on tall stems that are good for picking, with a vase life of four days. It has pretty good disease resistance, but only faint scent.

9 *R.* 'Roseraie de l'Hay' (Rugosa)
An exceptionally healthy, pristine rose often used for hedging as it's quick-growing and vigorous. Its loose, some would say messy, flowers are a deep magenta pink and do not pick well, but in the garden it has one of the most powerful scents of any rose.

10 *R.* 'Summer Song' (Shrub)
A warm amber-orange that glows. Good scent, long-flowering and healthy.

11 *R.* 'Tuscany Superb' (Gallica)
Like 'Charles de Mills', this is a classic old rose with a truly luscious colour and texture, dark crimson petals contrasting with the golden anther heart. This is not repeat-flowering, but has handsome foliage, so if well-trained it retains a garden presence even without flowers.

12 *R.* 'Warm Welcome' (Miniature Climber)
One of my favourite rich orange roses, invaluable in contrast to other rich colours, with quite upright growth. It flowers for a good five-month stretch from May into October. Good disease resistance, but little scent.

Warm

1 *Rosa* 'Belle Epoque'
(Hybrid Tea)
With all the colours of a
nectarine and a scent that
stops you in your tracks, this is
spectacular from the moment
it opens its first flower to the
last, which is often, incredibly,
in January. Good vase life.

2 *R.* 'Buff Beauty' (Hybrid Musk)
One of the best Musk roses in a
beautiful apricot-cream. Lovely
scent and good for picking. It
can be grown as a short climber.

3 *R.* 'Grace' (Shrub)
One of my new-found favourites.
It reminds me of a fruit salad in
look and perfume: fruity, pretty
and tempting. Very floriferous.

4 *R.* 'Honey Dijon' (Hybrid Tea)
A bred-for-picking rose, with

a good vase life of five days.
It's almost thornless, disease-
resistant and long-flowering.
I love its unusual colour, but
it's an acquired taste.

5 *R.* 'Koko Loco' (Floribunda)
This is a strange and beautiful
colour, sort of milky coffee
touched with mauve. It is quite
new to us, but seems to be
stronger and more flowery than
the similar 'Julia's Rose', and
has been recommended to me
by lots of cut-flower growers.

6 *R.* × *odorata* 'Mutabilis' (China)
One of my favourite all-rounder
roses. It flowers from May
to November and its name
means 'changeable' in Latin:
its flowers open apricot and
deepen to pink and crimson,
with marvellous foliage colour
in spring and autumn. It takes a
while to get going and mustn't
be overcrowded as it grows.

7 *R.* 'Phyllis Bide' (Rambler)
I've worked with Jonathan
Buckley (this book's
photographer) for over 20
years and this is one of his
favourite roses because it is
sweetly scented, flowers for
ages and is a beautiful mix of
colours. It's very healthy, too.

8 *R.* 'Pumpkin Patch'
(Floribunda)
I love the mix of colours you
get on a plant at one time –
deep golden, honey gold and
terracotta, fading to milky
coffee. This flowers prolifically
pretty much straight away
and over a long season. Good
disease resistance and vase life
of five days.

9 *R.* 'Tea Clipper' (Climber)
This only flowers once, but does
so prolifically and early (from
May) with a beautiful mix of
colours. Very healthy.

Our rose-for-picking trial with lilies, lavender, fennel, salvias, gaura and *Chasmanthium latifolium* scattered through.

Cool

1 *Rosa* 'Ferdinand Pichard'
(Bourbon)
A truly glamorous pink rose,
striped with purple. Huge,
delicious flowers are produced
over several months with a
good scent. This is one of my
favourites for putting beside my
bed. Pretty healthy, too.

2 *R.* 'Gertrude Jekyll' (Shrub)
A true, clear pink with a
deeper rosy heart. This excels,
not just for the vigour of the
bush and its repeat-flowering
(bushes start in May and should
continue to flower until late
October), but because once cut,
the flowers last for five days in
the house. It also has fantastic
scent. The only down side: the
stems are extremely prickly.

3 *R.* 'Ispahan' (Damask)
A beautiful mid-pink rose with
the RHS Award of Garden Merit,
and rightly so. It's a healthy
variety and it cuts well. We train
ours onto hazel frames, planting
three bushes to one dome. This
has Damask fragrance and a vase
life of three days.

4 *R.* 'Madame Alfred Carrière'
(Noisette)
Pure white to ivory, with a
touch of pink, this is ideal as a
climber for a north-facing spot.
Excellent for picking, good
disease resistance and scent.
The one planted in 1930 by Vita
Sackville-West on the front
wall of the South Cottage at
Sissinghurst is still there.

5 *R.* 'Queen of Denmark' (Shrub)
A famously reliable, soft pink
garden and picking rose that is
pretty prolific, healthy and has
good scent and vase life.

6 *R.* 'Starlight Symphony'
(Climbing Shrub)
A new rose made to look like it's
always been around, or might
be found in a hedgerow. This
one flowers for five months, has
good scent, excellent disease
resistance, is very prolific and
lasts well when cut.

7 *R.* 'Susan Williams-Ellis'
(Shrub)
A pure white rose that has a
hugely long flowering season.
It stands out, as there aren't
many other roses still flowering
in November here – and the
foliage is exceptionally healthy.
Fragrance is also good.

8 *R.* 'The Simple Life'
(Short Climber)
A more glamorous form of a
wild rose with single flowers
and beautiful anthers in peachy
pink. We've planted lots of
this in our soft-colour themed

Farmhouse Garden. It's trained, not as a climber, but bent down onto hazel hoops. It has really healthy foliage and flowers for ages, even sporadically through winter until it's pruned. In the winter it's good for picking, and has nice fat hips, but oddly, it drops its petals fast if picked in summer.

Salvias

I have included compact and medium-sized salvias in this selection. Hybrids such as the famous *Salvia* 'Amistad' are too big to plant under the skirts of roses – that's where you want a companion plant, not a dominator that will drown the rose. For a selection of my favourite larger salvias, see page 310 in October instead.

All these salvias are hardy, but may struggle with the excessively wet winters that seem to be becoming the norm, so are safest planted in spring and then propagated in late summer and autumn to guarantee plants from year to year. That said, many seem to over-winter happily here in our sheltered south-facing rose borders. Propagating is easy (see page 267), so we tend to take cuttings just in case we have a harsh winter and store our new plants in unheated cold frames to go out, if necessary, in April or early May.

Below Salvias 'Blue Note' and 'Mirage Deep Purple' with *Lavandula dentata* in a large pot in the rose garden.
Opposite Rose 'Rhapsody in Blue' with salvia 'Nachtvlinder' helping to keep it blackspot-free.

Best of the salvias

Amongst the best value garden plants, flowering for five months in Sussex. They look fantastic as a health-giving understorey to roses in a sunny spot, and drought-resistant, they're also excellent for pots. We've grown the varieties selected here for years, and each and every one is easy to look after, brilliant for pollinators, delicate, velvety and aromatic – what more can one say?

1 *Salvia* 'Blue Note'
A clear royal blue. We love growing this in a pot with one of the long-performing lavenders such as *Lavandula multifida* 'Spanish Eyes'.

2 *S. greggii* 'Stormy Pink'
I love the smoky, calamine lotion pink of this variety, along with the newly bred 'Mirage Soft Pink'.

3 *S.* 'Jezebel'
Another good red, which has slightly smaller flowers than 'Wendy's Surprise', but they stand out strongly from dark crimson calyces and stems.

4 *S. microphylla* 'Cerro Potosí'
A little taller than the salvias above and more branched, this is a rich deep pink. We've got this with the roses, but also in a pot with *Nicotiana* 'Lime Green'. In previous years – like our roses – this green-flowered plant has been clobbered by mildew, but not when it's interplanted with this salvia. It remains utterly pristine and flowers away until October.

5 *S.* 'Mirage Cream'
A lovely ivory, with a slightly apricot tone. We grew this for the first time in 2019 and loved it. It flowers from June to November.

6 *S.* 'Mirage Deep Purple'
Beautiful purple, one step brighter than 'Nachtvlinder', but equally long-flowering and valuable. We love this and now grow it not just with the roses, but in large containers too, as it looks good for so long with minimal TLC.

7 *S.* 'Nachtvlinder'
This is my long-standing favourite in luscious purple-crimson. There is now another purple salvia with a flash of white in its throat called 'Amethyst Lips', bred from the scarlet and white 'Hot Lips'. It's showier, but I don't like it quite as much.

8 *S.* 'Royal Bumble'
Brilliant red, velvety flowers contrast with the purple-black calyx. Hugely long-flowering and handsome.

9 *S.* 'Tutti Frutti'
A pretty coral-pink and apricot. This is similar in colour to the slightly larger-flowered, more salmon pink, *S.* × *jamensis* 'Señorita Leah'.

Our large Oast Garden pots in June with *Salvia microphylla* 'Cerro Potosí', *S.* 'Nachtvlinder' and *S.* 'Jezebel', which flower pretty well until November.

Practical
June

With most of our planting out completed by this time of year, June starts to feel a bit calmer in the garden. There are always odds and ends still to be planted, but we try to clear the polytunnel before we start sowing again. The main job for June is getting ready for the following year and sowing our biennials – and the sooner we can manage this the better.

From our trials here, we've found biennials benefit from being sown in June, rather than leaving it until July. And we often do another sowing of antirrhinums to give us flowers late in the year (see page 55). We also use this time to make plant teas to help keep the plants disease-free and organic. June is a good moment for brewing up comfrey, nettle and chive tonics, as all the plant ingredients are growing at full tilt. You can cut them to the ground and they'll grow back again in a week.

It's also weeding time. Lots of annual weeds are coming up to flower very quickly and they mustn't be allowed! We choose a hot, sunny morning, ideally after a night of rain (so the roots come up easily), and get stuck in using hands, trowels and hoes. Having a plastic sheet on the ground next to you where you can plonk all the weeds, plus a barrow, saves mess and tidying up later on. If we can see from germination rates that a particular area of the garden is full of annual weed seeds, we mulch heavily after weeding. The key thing with hoeing is little and often: do it before you even see many weeds and cut them off just after germination stage.

We start watering our pots now in earnest (see page 264). If it's particularly dry, water in the evening, not the morning. This avoids excess evaporation under the day's sunshine. There's also lifting and dividing any narcissi clumps that have become congested – left to their own devices, they'll stop flowering so well, and are best split and replanted straight away.

Our final job is hardly a hardship – picking flowers, particularly sweet peas, which are growing madly now.

Sowing Biennials

We don't direct sow seeds very often at Perch Hill, as we cram everywhere so full that any biennials sown this way tend to get smothered, and generally seeds don't germinate very well on our heavy clay soil. If you have lovely friable soil you can direct sow the tougher biennials such as wallflowers, sweet rocket and foxgloves this way in May or June (see page 127 for direct sowing).

Sowing biennials in seed trays is much the same as any seed-planting – for advice on this, see page 80. There are, however, a few things to look out for.

Sowing biennials with tiny seeds

❀ As a general rule, with almost any sowing, you should think of your seeds as gold dust and individually place all but the very tiniest (it'll save you time later on when pricking out). With truly dust-like biennial seeds, such as foxgloves, you just can't do that. For these,

Sowing tiny Iceland poppy seeds into a gutter.

we use deep and large trays or polystyrene crates (three times the size of a usual seed tray) for sowing. You should be able to get containers from a greengrocer or fishmonger.

❧ To sow, don't pour straight from the corner of the packet, but put a small amount of seed into the palm of your hand. Then scatter the seeds as far apart as you can over this wider-than-usual surface area. This is best done fast, in a quick swoosh of the arm, and from a height. Do the first swoosh at one end of the tray, the next over the middle and the final over the other end, trying not to scatter over the same area.

❧ Wide sowing means you hopefully won't need to prick out. With the increased depth of compost in the deeper tray and the seedlings well-spaced, you can skip that stage.

❧ Plant straight out in the garden from here.

Sowing biennials in guttering
Planting seeds in guttering is a favoured system at Perch Hill, one reason being that black plastic guttering absorbs warmth and provides a bit of bottom heat for rapid germination and quick root growth. It's particularly good for some of our biennials: Iceland poppies, for example, like many in the poppy family, don't like root disturbance, so this is the method for them as the seedlings can be slid straight into their growing holes when they're ready to be planted out. For full instructions on planting in guttering, see page 82 in March – but here are a few pointers.

❧ Fill your gutters with any old compost, homemade or shop-bought (the seedlings won't be hanging around in it for long).

❧ With biennials that don't like root disturbance, such as Iceland poppies, sow twice the number of seeds to plants you want to end up with. Even with this extra TLC, we get about 50 percent losses with this particular poppy.

❧ Once the seed has geminated, if you have two seedlings growing within 2-3cm (1in) of each other, thin out by removing one of the plants.

❧ Once germinated, cover brassica family

biennials, such as dame's violet (*Hesperis matronalis*) and wallflowers, with fleece to protect against flea beetle, which may pepper their leaves with lots of tiny holes, reducing some to lace. Once seedlings are 2 x 2cm (¾ x ¾in) plant out.

Making Tea Tonics

We have made comfrey and nettle teas here for a while, the first rich in potash, the second in nitrogen. You can use them separately or mix them together for a more broad spectrum feed.

Potash is necessary for good root, flower and fruit formation. We use it to feed the tomatoes and sweet peas every ten days once they've come into flower. And we use it to reactivate our compost heap, watering the heap well first and then sprinkling on the comfrey juice (in its concentrated form, undiluted) from a watering can.

We use nitrogen as a foliar feed for our pots, particularly with hungry plants such as dahlias. This is done once a fortnight starting about four to six weeks after planting, once the compost's

own nutrients start to become depleted.

In 2020, we added chive tea to our preparations. This contains sulphur – a natural fungicide – in its mineral make-up, and like the salvias planted with the roses (see page 171), chive tea helps to keep plants fungal-free. It has turned out to be useful for keeping mildew at bay on our sweet peas, courgettes and dahlias. It's too late to treat once the plants have mildew, but a fortnightly dousing is effective as prevention. Do this once the plants start flowering properly: with sweet peas, usually early June, and with courgettes and dahlias, aim for end of July.

Comfrey, nettle or chive tonic

❋   Cut the plant back to the ground and gather up all the stems and leaves in a trug or wheelbarrow.

❋   Chop everything up (you can include the flowers) and pile it into a box, bucket or water butt. Ideally, your container will have a tap in the base (such as a water butt), so you can extract the liquid, and a lid is useful as the mixture starts to smell as it breaks down, but any watertight container is fine.

❋   Fill the container with water, making sure to cover the plant material well. Leave the mixture to soak for 3-4 weeks.

❋   After this time, you can remove the liquid and store in bottles somewhere cool and dark. Alternatively, just take water from the mixture as and when you need it, always topping up a little with water and more plant material.

❋   To use the tea tonic, dilute with water to a ratio of about 1 part tonic to 10 parts water, then apply to the plants.

Staking

It's time to give April and May-planted hardy and half-hardy annuals, as well as dahlias and chrysanthemums, a bit of attention. The advice below depends on the plant being staked.

For heavy-stemmed, top-heavy plants with large flowers, such as the taller varieties of sunflowers, tithonia, *Nicotiana sylvestris* and ammi, it's best to stake each plant individually with a stake and string.

❋   We use willow or hazel sticks knocked firmly into the ground. Place them 5cm (2in) away from the bottom of the main stem.

❋   We use flexi-tie for most of our staking, as it is gentler on the plants and has a bit of elasticity and give, which prevents breakages. It's also easy to undo and reuse. We have a bag of previously cut lengths in the potting shed that I take out with me around the garden.

❋   Loosely attach the plant to the stake with a clove hitch. This knot is ideal for staking, as once attached, it stays in place and never slips down to the ground. A normal granny knot often does (see page 196).

❋   Don't pull the stem tight up to the stake, or it will snap in even a light breeze. The figure

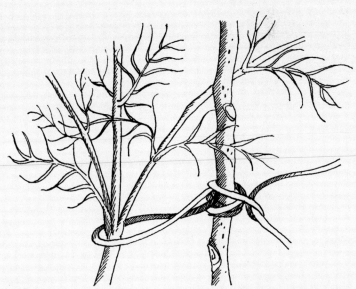

A clove hitch is a knot worth learning: start with a forward loop of twine or flexi-tie, right over left, followed by another forward loop next door, which then looks like two rabbit ears. You then tuck ear two behind ear one – just tuck, do not rotate them. That double loop then goes over your stake and you twist the longer end of twine onto the plant with a figure of eight, looping it back onto the stake.

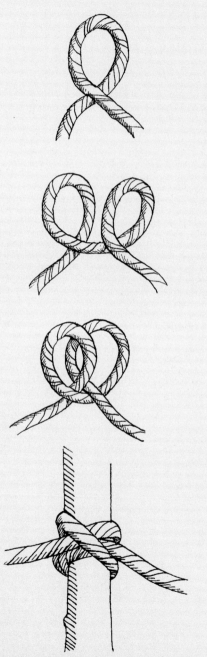

of eight means that there's a layer of string or flexi-tie to act as a cushion between the stem and stake, it also gives more support than a straight loop.

For bushy, chunky plants such as dahlias and chrysanthemums, it's better to have a cage of three or four stakes, rather than just a single stake. This supports them well and they'll really need that from July onwards, when most reach to well over a metre (3ft) and catch the wind like a sail. It feels way over the top when you first do it – our dahlia garden looks like plant prison – but quickly, everything grows and the support system is hidden.

❧ Check the final height of the variety you're growing and support them with a triangle or square of uprights. Wind string at both one-third and two-thirds of their final height.

For more delicate, lighter plants that are tall, such as snapdragons and larkspur, as well as shorter plants that tend to flop over, such as cerinthe and the *Phlox drummondii* varieties 'Crème Brûlée' and 'Cherry Caramel', we use jute netting.

❧ Support the plant with at least one layer of jute netting, stretched horizontally and taught between willow stakes or bamboo canes.
❧ Get this support system in place as soon as you can so that the plants, once settled, grow up through it and are supported early on.

A double layer of jute netting supporting weak-stemmed phlox 'Cherry Caramel'.

❀ I find phlox do best with a double layer of jute: one tied very low at 20cm (8in), the next at about 40cm (16in). This results in pretty straight stems for picking and the rows look orderly in the garden.

For our bulb lasagne beds that are permanent (see page 74), we have a wooden grid structure made from hazel that supports groves of dahlias underplanted with narcissi. It lasts for about five years before it needs replacing.

Dividing Narcissi

This is the ideal time to lift narcissi and divide them as they are in the green and just going over. Transplanted like this, still in active growth, their roots settle in quickly to their new home. This method is the same for snowdrops and aconites, though these should be divided in April.

❀ Deadhead any remaining flowers: you don't want the bulbs having to put energy into making seed, as they need to concentrate on roots.
❀ Lift a whole clump in one go, getting well beneath the bulbs with a spade. Make sure you dig good and deep so you don't slice into any bulbs.
❀ Place the clump on a tarpaulin and divide the bulbs into groups of three to five.
❀ Improve the soil in their new home with organic matter. Replant to the same level.
❀ Never do that traditional thing of tying their leaves in knots. They need to be able to photosynthesize, so leave the foliage as it is to die back completely and help ensure good flowering the following year.

July

The garden is jam-packed in July. All the hardy annuals – sweet peas, calendulas, ammi and scabious – are still flowering. Most of the biennials are tailing off a bit, but foxgloves and sweet Williams are still looking good. These are now joined by the first dahlias and half-hardy annual flowers. It's as though the tide hasn't quite gone out on one whole tranche of plants, when another comes in.

As in June, the whole place is incredibly fragrant. Sweet peas are at their best, right to the top of their teepees and joining up over their arches, and most of the roses are still flowering. We've recently found a whole bunch of newly bred antirrhinums, including the Chantilly series and another variety from America called 'Apple Blossom'. They are all excellent, healthy and long-flowering with tall stems, and each has an unusual scent – Chantilly is fruity, 'Apple Blossom' spicy like cinnamon. One of my favourite soft colour mixes for this time of year is *Antirrhinum* 'Chantilly White', *A.* 'White Giant', which self-sows here, with *A.* 'Apple Blossom' scattered through both for gentle colour contrast.

Nicotianas are also flowering now. Dusk to dawn, *Nicotiana alata* 'Grandiflora' fills its dark corner with the most delicious perfume. I love it underneath our bedroom window.

We grow lots of jasmine and honeysuckle, planting them wherever we can on hedges, fences and walls. I like the new shorter honeysuckles (such as 'Chic et Choc') and buddleja (including 'Dreaming Lavender'), which fit happily in a pot. They have been bred to flower for two or three times as long as their wild forms, but like their forebears, have great scent.

The Oast garden with pots of agapanthus 'Midnight Star', dahlia 'Hotcakes', *Salvia viridis* blue-flowered, *Tagetes* 'Burning Embers' and clematis 'Madame Julia Correvon'.

One of the first things I planted at Perch Hill were three star jasmines (*Trachelospermum jasminoides*) on the posts supporting the balcony overlooking the Oast Garden. This evergreen semi-climber is not fully hardy, and away from the microclimate of London or a southern city, needs the protection of a building. Even so, we lost one of them in the cold snap one year, but the other two seem fine. They grow slowly, but, twenty years on, have crept up to form a fragrant curtain around the edges of the balcony on the first floor.

Apart from jasmine plants, the colour white and anything partly white is banned from the Oast. It's the intensely coloured varieties only here, and phlox 'Blue Paradise' fits exactly. It has a strange, subtle colour, sometimes blue and at other times mauve, with a scent of honey mixed with a pinch of garam masala. It produces this in massive bursts, particularly early in the morning. Lots of phlox are moth-pollinated, this one included, with dawn being its most fragrant time.

When I started the garden, the two plants I was obsessed with growing were rosemary and lavender. I didn't know at the time that neither would thrive on the heavy Perch Hill soil.

Below Crocosmia 'Lucifer' and phlox 'Blue Paradise' in the Oast Garden.
Opposite Looking from the Dutch Yard into the Oast Garden in July, with a giant rocket of clematis 'Madame Julia Correvon' in full flower.

I lost them time and again, until I learnt they need incredibly effective drainage. They'd probably both be happier growing in builders' rubble than the rich, heavy, water-retaining soil here – just look at them on the edge of a terraced Mediterranean vineyard or olive grove and you'll see what I mean. We do grow lots of both here now, but wherever they are, you can be sure barrow-loads of grit have gone into their planting holes beforehand, and even with this, they do best on a slope.

The July scent is not just for our pleasure. Until recently (before my husband Adam became dangerously allergic), we had beehives, and at this time of year the harvests would be huge. Our beekeeper says she could pretty much expect a honey crop three times' the usual volume in July. I love the idea that our flowery, fragrant dot on the map is not just keeping her bees here, but may also be feeding thousands of wild bees, butterflies and hoverflies. That makes the place feel worthwhile on another level.

Opposite *Lavandula angustifolia* 'Munstead' and 'Hidcote' lining the path of our edible garden.
Below *Phlox paniculata* 'Jeana', 'David' and 'Blue Paradise', with *Stipa gigantea* and our echinacea for picking trial, including 'Summer Salsa', 'Summer Samba', 'Marmalade' and 'Eccentric'.

Sweet peas

I was on a bike ride with Adam in the south of Italy a few years ago, just south of Matera. We were there for a bit of sun and freedom after completing a garden at the Chelsea Flower Show. The caper vines were like flowery wigs on all the walls, the verges were full of wild carrot, vetches, scraggy self-sown figs and sainfoin, and the olive groves had the odd splotch of colour from the scarlet-orange flowers of pomegranate trees. But what I was determined to find was the original sweet pea.

I imagined hedgerows swathed in billowing vines of scented peas, like a garden in the wild. But when I found the precious pea, it wasn't like that at all – it was just three rather scrappy plants in one spot, but I was still thrilled to have found them.

All the sweet peas flowering in our gardens today descend from this wild plant, brought into cultivation by a monk, Francis Cupani, in Sicily in the late 17th century. The varieties we know as 'Cupani' and 'Matucana' are like the wild species but bigger and showier. They are both bi-colours (purple and crimson), and good, floriferous forms with indisputably the strongest perfume. We grow more 'Matucana' than any other sweet pea at Perch Hill.

There was breeding work done on the wild pea at Chelsea Physic Garden and a few other places around the world, but after a nearly two-hundred year gap, Henry Eckford took on the sweet pea-breeding mantle. In 1888, he set up his own nursery in Wem, Shropshire and dedicated the latter years of his life to improving and expanding a small band of sweet peas.

Eckford increased the flower size of 'Cupani' and expanded the colour range, creating what we now call the Grandiflora. Compared to modern varieties, these have relatively small flowers and short stems, but many are irresistible. We grow lots of Eckford's varieties here, including the zingy pink bi-colour, 'Prince Edward of York', 'Black Knight' (crimson) and 'Lord Nelson' (navy), which I'd miss enormously if they weren't with us in July.

Sweet pea breeding was then taken up by Silas Cole, who was the head gardener at Althorp Park, Northamptonshire, the

Picking sweet peas including a few long vines and tendrils to add shape when arranging in a vase.

country seat of Earl Spencer. Cole grew and loved lots of Henry Eckford's varieties and was triumphant in 1900 when one of his plants threw up a 'sport' (a chance seedling) with even bigger flowers and a distinctly wavy petal edge. A similar frilly form was also found in the same year by William Unwin, a grower of sweet peas for the Covent Garden flower market. This became the basis of the Unwins Seed Company.

Everyone went mad for this flamboyant scented flower. The ivory and pink 'Anniversary', deep pink 'Barry Dare' and rich maroon 'Winston Churchill' are all Spencer types, although bred more recently. And I love 'Blue Velvet', which just keeps flowering. This length of performance in a sweet pea is a real breakthrough. It makes it a genuinely useful border plant, as well as a scented cut flower. We had 'Blue Velvet' flowering last year surrounded by the late-season bulb, *Galtonia viridiflora* and the towering perennial, *Veronicastrum virginicum* 'Fascination', which are two August (rather than July) favourites.

Since then, we've added lots of new ones, including the Modern Grandiflora varieties developed to have large flowers and lots of them, with the long stems of the Spencer types and the fantastic

Harvesting from the soft and warm palette (see p218). I secure sweet peas with elastic bands as I pick.

scent of 'Cupani'. I love 'Almost Black' from this lot, a near-black, glamorous and highly scented variety. When I want a softer colour, I often choose the perfectly pretty 'Fire and Ice'.

Frustratingly, you can't count on the availability of a particular sweet pea variety. They are prone to crop failure and almost total wipe-out, and this has driven me mad in the past, when I couldn't get hold of a long-standing favourite. 'Painted Lady' is a good example: it was my second favourite after 'Matucana', early to flower with a delicious scent, pretty colour and healthy for twenty years or so, but its germination can be sporadic. We've been trialling new ones hoping to replace it, but have yet to find anything quite as good.

Recently, I've come to the decision that it doesn't really matter. It's the palettes and colour families that are key, and as long as I can get at least one form in each of the key colours, even if it's not the same named variety, it's alright. Rather like tulips, I think lines can get exhausted and out-done by new hybrids, so I now don't mourn the old varieties as much as I used to. When looking for new ones, I think the RHS Award of Garden Merit is worth considering (look at 'Valerie Harrod' and 'King Edward VII').

Here's a tip for picking, particularly the shorter-stemmed, old-fashioned types whose stems get shorter as the season goes on. They can get so short, in fact, that as you pick them, stem-by-laborious-stem, they're forever falling out of the container you're putting them into (and then the vase you arrange them in). I do one of two things so I don't get fed-up with these pygmies. Either I pick them as they are, short-stemmed, and use elastic bands on my wrist: once I have a fistful, I bunch them with a band and put them into a bucket of water. I then arrange them in the vase like this, and only when their stems are held by the neck of the vase do I cut the elastic band. It works brilliantly and saves tons of fiddle and time.

The other way is to forget picking sweet peas stem by stem, and instead go for a main stem, cutting the top section with three or four individual stems and flowers along that length, twirly bits and all. As long as you cut above a pair of leaves, more axillary buds will form to replace the leader, which keeps the plants flowering longer. Then I trail the vine in amongst other flowers as I arrange them, in the style of a Dutch still life painting. It gives my arrangements a much more interesting shape and keeps the plants going longer, without the need for endless deadheading.

A mix of old and new sweet peas from our trials, including 'Winston Churchill', 'Princess Elizabeth', 'King Edward VII', 'Linda Carole', 'High Scent', 'Beaujolais' and 'Mollie Rilstone'. They're underplanted with a snapdragon mix, *Antirrhinum majus* 'Brighton Rock'.

In terms of planting our favourite varieties, colour segregation is the key. There is something triumphantly lovely about a total colour mix, as you see in summer allotments, with reds next to whites, purples, pinks and blues. But at home I'd rather refine the colour range and think carefully about which colours I want.

For me, it works best having the rich and brilliant hues in one place, the soft and warm in another and the cool shades in a different spot. Jammed up together these palettes compete and detract from each other, but kept apart they are one of the great sights of our summer garden.

Below Sweet pea 'Flora Norton' in the Farmhouse Garden with the olive behind.
Opposite A vase of soft and warm sweet peas including 'Apricot Sprite', 'Princess Elizabeth', 'Anniversary' and 'Mollie Rilstone'.

Best of the sweet peas

*We grow a mix of historic sweet pea varieties
(such as 'Matucana') and brand-new ones, which
I select when I visit breeders every couple of years.*

Rich and brilliant

I like mixing the dark, rich varieties, such as the blues and crimsons, with the saturated bright forms in deep pink and red. The bright ones together are too clashy and vie for attention, while the dark ones get a little sombre on their own. Mixed, these two palettes really help each other out.

1 *Lathyrus odoratus* 'Almost Black'
One of the new so-called Modern Grandiflora, with good scent, longer than average stem length and more (quite small) flowers than usual on one stem. The key thing about this one, though, is its colour. In our trials of the dark and rich, this is the darkest, the nearest to black we've ever grown. 'Black Knight' is a touch less dark, but has stronger perfume.

2 *L.* 'Bouquet Crimson'
A Multiflora introduced ten years ago by Roger Parsons, who is one of the contemporary heirs to the great succession of breeders. Most sweet peas in my garden have three or four flowerheads to a stem, but these have five or more, and Roger says he's counted a few with up to twelve. This means huge impact, but also that each stem gives you a longer show in the garden and in the vase. The bottom bloom might be fading after a few days, but the upper ones will just be starting to open, so it has almost double the average vase life. Anything with 'Bouquet' in its name hails from this group.

3 *L.* 'King Edward VII'
A strong, deep red, this is one of my favourite colours of sweet pea as I find it goes well with bright and dark colours, and it stops the soft apricots and pinks from being too sickly sweet. It's similar in tone to the slightly less perfumed 'Barry Dare' and 'Winston Churchill'.

4 *L.* 'Lord Nelson'
A variety I've loved and grown here for twenty years with top-notch scent. I like 'Top Hat' too, in a fractionally richer tone, and love 'Blue Velvet', with slightly less scent, but a hugely long flowering season.

5 *L.* 'Matucana'
The sweet pea classic, against which we measure everything else – particularly scent – in all our sweet pea trials. I would never fail to grow this variety; no summer garden is complete without it.

6 *L.* 'Prince Edward of York'
I love the perkiness of this bi-colour pink, bred directly from 'Matucana' and with good scent.

Soft and cool

I love all the cool colours together – the blue-mauves, those in the Shifters group (such as 'Blue Shift' and 'Turquoise Lagoon'), as well as the ivories outlined or stippled with purple, along with the stark whites. They are all lovely growing and arranged in a group.

1 *Lathyrus odoratus* 'Aphrodite'
Introduced by Unwins, with up to fifteen flowers per stem, this is a beautiful pure white. Perfect for summery wedding bouquets.

2 *L.* 'Erewhon'
This is an unusual colour, a sort of smoky mauve bi-colour. It flowers early, with long stems and good scent. It's a new-found favourite.

3 *L.* 'Fire and Ice'
Unusual and pretty, with a white base, lightly stippled pale pink in the wings and bluey lavender in the nose. It also has good scent and a sort of translucent delicacy.

4 *L.* 'Nimbus'
A new-found favourite for being very early and long-flowering. From an autumn sowing, this will be in full flower in my garden a week before the Chelsea Flower Show in late May. I'm able to pick many bunches of this – and only this – to take up to our stand. We have it underplanted in the garden with the similarly early-flowering honeywort (*Cerinthe major* 'Purpurascens'). They make a great garden and vase pairing.

5 *L.* 'Sir Jimmy Shand'
Lovely, generous large flowers with wavy petals, good scent and a pretty colour – this is a quintessentially romantic sweet pea variety.

6 *L.* 'Turquoise Lagoon'
A variety from the Shifters group, which open one colour and turn bluer as the flowers develop. 'Turquoise Lagoon' opens mauve and goes turquoise-blue as the flowers age on the plant. With this colour variation, I love it simply in a vase on its own.

Soft and warm

I think the soft, romantic colours – peach, apricot, gentle pink and cream – are tricky to find in sweet peas, but wonderful planted together. They are improved by a splash of deep red or crimson to tone down the potential for over-sugary sweetness. We often use 'King Edward VII', 'Henry Thomas', 'Pip's Maroon' and/ or 'Winston Churchill' in this role.

1 *Lathyrus odoratus* 'Anniversary'
Similar colour to another lovely variety, 'Mollie Rilstone', this has petals of ivory with a pink shading to the edge. Almost interchangeable, 'Anniversary' has an exceptionally long flowering season, 'Mollie Rilstone' better scent.

2 *L.* 'Apricot Sprite'
An unusual coral and pink sweet pea with good scent. This is probably my all-round favourite of this colour palette.

3 *L.* 'Linda Carole'
A brilliant new sweet pea in a deep rich pink as if pigment has been dusted over a white base. Good scent.

4 *L.* 'Maloy'
With flowers in an orange-pink, this is another Multiflora (like 'Bouquet Crimson'), but usually with six blooms to a stem. This colour is an acquired taste and a little too salmony for me, but it is a very floriferous variety.

5 *L.* 'Mrs Collier'
A warm ivory-cream, with neat flowers and lots of them. Good scent.

6 *L.* 'Princess Elizabeth'
A soft shell-pink Spencer type with good scent and decent stem length.

7 *L.* 'Valerie Harrod'
Deep rose pink with a softer petal centre that stops this being too much. There are long stems and lots of flowerheads on this newly bred cut flower form. This has been awarded an RHS Award of Garden Merit.

Nicotiana

Have you ever witnessed the crazy bun fight-cum-jumble sale that is the last couple of hours on the final day of the Chelsea Flower Show? Lots of stands sell off their plants very cheaply, or give them away for free, and you see people leaving the grounds struggling under the weight of climbing roses, in-flower dahlias, armfuls of alliums and foxglove 'Sutton's Apricot' swaying about – it's like a great end-of-festival parade.

The first time I went to the show over 30 years ago, Jane Fearnley-Whittingstall had designed a garden full of roses 'Charles de Mills', 'Tuscany Superb' and 'Evelyn', together with tobacco plants in between. When the show ended, Jane gave me five huge *Nicotiana sylvestris*, a plant I'd never previously seen. I managed to get them on the tube and back home and I planted them that evening in my jungle-like but rather lovely back garden.

I planted all five in one place next to a cardoon. They had the architecture and substance to hold their own. Even though they had been forced into flower two months earlier than usual, their great, white, scented towers went on until October.

When I came to tidy and cut back the garden the following spring, I found that where the tobacco plants had been in the skirts of the cardoon, there was a carpet of brilliant green. Half-hardy in most places, in the microclimate of a southern city garden, the plants' self-sown seedlings had survived the winter. With no need for sowing, I had them in the garden every year without fail.

Now we grow the same *N. sylvestris* in big clumps between *Acanthus mollis* 'Rue Ledan' on the north side of the Perch Hill school. Not only is this one the biggest and boldest annuals, which rockets up to 1.2m (4ft) within weeks, but it is just as good in shade.

I always think you can get away with planting plenty of shrubs in shade. Their growth curve is slow and they can photosynthesize at a low rate in low light levels and still do fine. The reverse is true of almost all annuals. They need to grow super-fast and get

Nicotiana 'Whisper Mixed' with the beautiful and delicate grass, *Phaenosperma globosa* in one of our large pots in the shade on the north side of the barn.

up to flower in order to form seed in only a few weeks. With this steep growth curve comes a need for good light. We plant all our annuals here in a sheltered position and in full sun. That seems to be the sensible path to success if we don't want them to become etiolated, their necks reaching out for the light in a skinny, pallid, rather unhealthy looking way.

But not so with nicotiana. They are part of the nightshade (*Solanaceae*) family, which might explain why, particularly *N. sylvestris*, *N. mutabilis* and *N. alata* 'Grandiflora' thrive in partial or dappled shade. I've always thought that the huge surface area of the leaves of *N. sylvestris* must help too, those elephant ears must enable it to make lots of food, even in shade.

The one thing worth knowing about them is that they love a moist soil – another reason they probably thrive in shade. In a dry summer, they are prone to mildew, so water their roots well. If you see any sign of mildew, or better still anticipate it in a drought, give them a good dousing in chive tea (see page 193) every three or four days until it rains well.

Below *Nicotiana* 'Lime Green' with *Salvia viridis* blue-flowered and ageratum 'Blue Horizon' in the Annual Cutting Garden.
Opposite top *Nicotiana alata* 'Grandiflora' in the shade outside my office door; it smells fantastic.
Opposite bottom Our shaded, white border in July, with *Nicotiana sylvestris*, Japanese anemones, (*Anemone* × *hybrida* 'Honorine Jobert'), ivy 'Green Ripple' and *Acanthus mollis* 'Rue Ledan'.

Best of the nicotiana

This is one of my favourite families of half-hardy annual plants, some delicate, like hovering butterflies on a stem, others statuesque and huge; many have scent.

1 *Nicotiana alata* 'Grandiflora'
With an exotic, stephanotis-like scent, this is an exceptional plant, but it needs shade. I was growing this for years in sun and it just collapses, but when I crammed a pot with several plugs and grew it in an east-facing corner it did much better. When the sun was on it in the middle of the day, it looked half-dead, but as soon as the temperature cooled by 4ish, it perked up and looked and smelled marvellous. But it also hates drenching rain, when it collapses rather pungently (not in a nice way).

2 *N. alata* 'Lime Green'
A sharp lime that mixes so well with blues and purple. We often grow it as a cut flower alongside *Salvia viridis* blue-flowered and *Ageratum houstonianum* 'Blue Horizon' (see page 222) – the three together look handsome growing and arranged side by side. I also love this plant arranged with the deep red *N.* × *hybrida* 'Baby Bella' and the green and red *N.* 'Tinkerbell'.

3 *N.* × *hybrida* 'Baby Bella'
A rich crimson plant, its flowers a great contrast to its bright green leaves. In full sun, in a dry year, this might need watering or a douse or two of chive tea, but it's worth it.

4 *N.* × *hybrida* 'Whisper Mixed'
Bred from *N. mutabilis*, this is similar, but flowers for even longer, and whatever the weather, seems less prone to rust or mildew. We grow this in a large pot on the north-facing side of the barn and it thrives, even in dappled shade.

5 *N. langsdorffii*
On our doorstep in July we have two pots filled with this and it does well, though it does collapse after a few days of solid rain.

6 *N. sylvestris*
Woodland tobacco is the classic shade-lover. Moth-pollinated, this pulses out scent late at night, so plant it near where you might have supper in the garden, or below a bedroom window.

7 *N. tabacum*
Common tobacco is a statuesque plant, like *N. sylvestris* in scale, but with pink tubular flowers. This is used as bedding in parks in France and looks magnificent. We should all grow it more as it gives huge impact from just a cheap packet of seed. A regular application of chive tea over its huge leaves is a good idea to keep them fungal-free.

Practical
July

July is a month we can really enjoy the garden at Perch Hill. There's stuff to do – not least lots of watering, particularly of our pots (see August) – but overall this is a quieter-than-usual month for us.

Our main tasks are picking (liveheading) flowers and then conditioning them on a daily basis to help them last as long as possible. With so many flowers out in the garden and temperatures on average pretty high, July is the month when we take time to condition almost every flower we pick. This makes a huge difference to maximizing their vase life.

We're also deadheading the flowers that we have failed to harvest and saving seedheads too, as well as bringing in allium globes while they're still pristine to dry for decorative use later in the year.

With sweet peas at full tilt, it's easy to think there's nothing more to do, but they're pretty high-maintenance plants and need regular TLC: watering in dry spells, tying in and also picking (ideally every other day) to keep the seedpods from forming.

Maintaining Sweet Peas

To keep sweet peas looking good for as long as possible, they need a bit of maintenance.

❀ As the plants grow, tie them in to whatever frame they're growing on, don't let them flop around. They'll grow more quickly and make stronger plants tied in regularly, once every ten days or so.

❀ You might hear that you need to pinch out all the curly tendrils because they take energy from the flowers and attach themselves to flower stems and bend them into curves. It's a lot of work, so I try to remove any I see while I pick, but I don't get bogged down in the task. There's a new breeding programme to develop tendril-less sweet peas. It's worth keeping an eye out for these.

❀ At this time of year, we pick, pick, pick. If we see any seedpods as we're cutting, we snip these off as well. You don't want plants forming seed as this will stop them producing flowers.

❀ To keep them growing strongly, we give our plants a high-potash feed fortnightly. Any tomato food will do, but in our trials at Perch Hill, homemade comfrey tea has been the most successful (see page 193).

❀ Water plants properly in a drought. Sweet peas need thorough watering. Let them dry out and the mildew will soon appear as they get stressed.

❀ In 2020, we did a trial in which we used chive tea (see page 193). This has a high sulphur content and is a good organic anti-fungicide. A fortnightly spray with this, in addition to the watering and feeding regime, meant we had pristine plants for a good three weeks longer.

❀ With unusual varieties, we sometimes harvest some sweet pea seeds in late July and August, before we take the plants out, leaving the roots in the ground (because they're nitrogen fixers). Most come true from seed (except F1s) because they are self-pollinating. Store the pods in a labelled envelope and leave

Collecting nigella and opium poppy seedheads to hang in the barn to dry. I like the feeling of abundance they give inside, even in the winter.

them to dry fully for about 4-6 weeks before you empty them into a labelled paper bag and then into a container to store in the fridge.

Harvesting Seedheads

It's a good idea to pick all your alliums as soon as they finish flowering, when they're still green and before they become brown and brittle. Otherwise, they can be forgotten and end up trashed by the weather. We then hang them from their stems upside down until December and use them as part of our Christmas decorations.

More durable, weather-resistant and robust are the seedheads of opium poppies, agapanthus, sea holly, *Smyrnium perfoliatum*, angelica and fennel, so bring some of these indoors to hang and dry at the same time. Later in the year, we add Chinese lantern, hydrangeas and Japanese anemones, which all have good shapes and durability when dried. We try to remember to do a monthly garden seedhead harvest, picking when plants are starting to dry, but still at their best. They then go in to hang with the rest.

Deadheading

As well as picking, we are pretty busy deadheading in the garden in July. We deadhead all our roses as often as we can, snipping off their browning heads to a bud or leaf below. This helps promote the formation of axillary buds and more flowers to follow.

All hardy annuals, and particularly scabious, benefit from picking and deadheading. With scabious, if it's covered in seedheads, the easiest way to deal with a clump or row is to get out the hedging shears and cut the whole line back by at least one third. If you do this early enough in its life cycle (in July), you'll find that within a couple of weeks, every plant is covered with buds and new flowers – and it takes only five minutes, compared to nearly an hour to deadhead individually.

Picking Cut Flowers

Picking flowers and arranging them was one of the main things that got me interested in gardening; one of my great pleasures is bringing the outside in.

To do this with minimal waste, I want to make sure that I pick and condition my flowers in the right way, to give them the longest possible vase life. The combination of growing the right cut-and-come-again plants and treating them in the right way will give you a guiltless harvest.

❧ We wear gloves for picking, crucial if picking euphorbias – their milky sap is highly allergenic. Having ended up in hospital with sap in my eye, I am now very careful.

❧ We don't pick in the heat of the day, but in the evening or first thing in the morning. Plant cells are full of water (turgid) after a period of darkness when levels of transpiration and photosynthesis are low. Turgid cells don't flop as easily.

❧ Cut the stems at an angle so they don't form a flat seal on the bottom of the bucket. This increases the surface area of xylem (the means of water transport).

❧ Put cut stems straight into cool water, not into a pretty basket. This makes a big difference to the vase life. When picking, I always take two buckets into the garden: one is a third full of water and one is empty (for stripped leaves). We use milk pails rather than florists' buckets, as these have handles.

❧ Even when they're in water, don't leave the flowers in direct sun. Find a spot in the shade as you continue picking.

❧ Remove the leaves from the bottom two thirds of the stem (at least) as you pick, stripping the leaves into the empty bucket. Pick, strip, plop, pick, strip, plop. This saves lots of time and mess in the house later. There are really three reasons for doing this. You don't want the leaves right in the middle of the arrangement, they just clog it up. And you don't want any leaves below water level as they'll rot and get in the way of the vase neck. Thirdly, with fewer leaves there will be reduced water demands on the stem, making it less likely to flop.

* When you pick annuals, biennials and dahlias, don't cut them to the ground. It's a waste of lots of potential flowers, and with some varieties, will kill them. Instead, take out the leading shoot, or the tip of a side shoot if you want a more delicate stem, cutting just above a side-branch or leaf with a bud, which will grow and produce more flowers.

* I bear in mind the size of the arrangement I'm picking for and aim to cut to that scale. If you cut a long stem and end up just using the top two thirds in the vase, you'll be cutting a third off and wasting lots of axillary buds below the cut (this is the case with annuals and dahlias in particular). Cut to the length you need and you'll avoid delaying re-flowering.

* When picking bulbs that have leaves up the flowering stem – such as tulips, *Fritillaria imperialis* and lilies – leave a portion of stem behind. If you steal too much of its photosynthetic food factory (its leaves), the bulb will not be able to produce enough food to survive and flower well the following year. This does not apply to alliums, narcissi and hyacinths, which all have leaves at the base and their flowering stems can be cut to the ground without affecting the bulb's chances of survival.

* With perennials and shrubs, it's especially important to pick sensitively: look at the plant to see how you could improve (or at least not damage) its shape. We don't pick all our stems in one place, but take one here, one from there. In that way, we're thinning out, not mowing flowers.

Conditioning Cut Flowers

Most plants we pick in summer get one or other of the conditioning methods outlined below, but there are a few exceptions and I've listed these first.

Tulips: See page 118 in April for more advice.

Lilies: To cut lilies for a vase, you need to remove their anthers as soon as the flower unfurls. If the anther is allowed to ripen on the flower, it will pollinate and fertilise it, meaning the flower has done its job and so dies. If you remove them, it will keep going for longer. Lily pollen is, anyway, a pain if it drops onto tablecloths and stains. And we're often told it can be poisonous to pets. You don't need to sear the stem ends in boiling water or submerge in water, as these techniques have little impact on lilies.

Sweet peas: In sweet pea season I make up a jar of sugar syrup: dissolve 300g of caster sugar to 150ml of water over a low heat. Wait until it cools, then bottle. A ratio of 2 parts sugar to 1 part water works if you want to make less. I add a drop of this sugar solution to our vases of sweet peas, and this makes them last another couple of days.

Searing stem ends

I have loved picking flowers since I was a young girl, but wasted so many from my parents' garden until my aunt showed me how she kept her hellebores from flopping. She boiled a pan of water on her Aga and stuck the stem ends into the pan, and then straight into a vase of cold water. It has become second nature to me, and before I head out into the garden to pick flowers, I always put the kettle on.

Searing stem ends is a hugely effective way to extend the vase life of a cut flower. The boiling water breaks down the stem's outer layer and increases the surface area of xylem (the means of water transport). This prevents flowers from flopping.

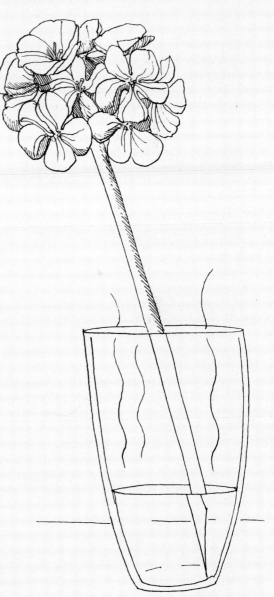

❀ Any plant that begins to look floppy once cut should have the ends of its stems seared in boiling water. I use the searing technique much more during spring and summer than I do in the autumn. Later in the year, the stem cell walls have more lignin in them and are woodier and more upstanding. In the spring, there is lots of fleshy new growth, which is more likely to drop. I even sear rosemary stem ends in spring, because if I don't the growth tips tend to bend and then don't look great. As a general rule, in spring and summer I sear 75 percent of what I pick, in autumn, only 25 percent.

❀ The time in the boiling water is proportional to the texture of the stem. Soft stems need just a few seconds, woody stems need more like 30 seconds. For example, bluebells should be seared for only 10 seconds, whereas lilac for 30. Don't leave soft stems in too long or they'll disintegrate.

A harvest of cut flowers for a family
wedding, sitting in water buckets
in the cool shade.

❦ The amount of stem you sear is proportional to the length of stem you've cut. I sear about 10 percent of the length – so if it's 1.8m (6ft) tall, sear 15cm (6in). If it's 45cm (18in), sear 2½-5cm (1-2in). With short-stemmed plants, take care to keep the flowerheads away from the steam. Enclose the flowers in a sheet of newspaper to protect them.

❦ To sear the stems, prepare two vessels. In one (usually a mug will do), pour 2½-5cm (1-2in) of boiling water (or the amount relevant to the length of your stem, see above). In the other (usually a vase or jug), add cold water. Plunge the stems into the hot water for the appropriate time and then straight into cold water.

Submerging

There are a few plants that benefit from submersion. We do this with plants that have complex structures with large surface areas. Hellebores (which are seared before submerging), hydrangeas, peonies and some shrubby foliage plants including smoke bush, dogwood and hostas.

❦ Float the stems and flowers in a sink or bath of tepid water, pushing them below water level (the flower as well as the stem) to start with. A few hours will do, but if you can, leave them overnight. Even if you've left them in the car, or in a basket in the garden, and they have seriously flopped, they'll fully recover in the bath. I used to worry about submerging white peonies, thinking the flowers would turn brown, but even delicate blooms like these really benefit from submersion.

Resting

Resting is relevant to all plants; everything picked here is rested before being arranged, which means placing stems in a bucket of water somewhere cool and dark, ideally for 12 hours, before arranging.

I've realised over the years that this is the most important step in transforming garden flowers into great cut flowers. Don't pick and plonk, but rather pick, then condition, then rest, and then arrange. Pick, condition, rest, arrange – that's the mantra here!

But bear in mind the temperature when you're choosing where to rest your flowers. Near a fire or radiator, you'll kill them quickly. Being kept cold, but not freezing, makes a huge difference to the vase life of any flower. So particularly in winter, with central heating often full-blast, flowers go on the doorstep (if it's not frosty) or at least into a cool room for 12 hours.

Flower Food

Adding flower food to the vase water is important. We tend to use an acidifying agent, such as vinegar or lemon juice, to reduce the proliferation of bacteria. You can use bleach, but we have a septic tank here, so try to avoid using it. By adding a weak acid, you create a hostile environment for bacteria. Bacteria create the slime at the stem ends that makes the vase water stink. It also blocks the stem and stops the plants taking up water.

In a 30cm (1ft) high vase, we add a good slurp (about five tablespoons) of clear, cheap distilled malt vinegar. This is the basis of the adage of adding an aspirin, or half a glass of lemonade, to your cut-flower water.

Arranging

Quite often I want a mixed arrangement or a hand-tied bunch, rather than a simple vase of just one or two different flowers. When that's the case, I follow this method. For most medium-sized or large mixed arrangements, I pick the following ingredients, which covers three or four foliage plants and three flowering ones.

Foliage
You'll need about seven stems of the primary, filler and upper-storey foliage plants, arranging those first. After you've added the flowers, add the vine.

Primary: This is the first plant I pick, and more often than not, it's *Euphorbia oblongata*, in brilliant acid-green. This is one of the best foliage plants you can grow as it's robust and upstanding, making it an ideal base for an arrangement. Some people don't like the vibrancy of its acid-green, but in fact this provides the bunch with a skeleton, rather than flesh, so you in the end you hardly see it.

Filler: Lots of plants can fulfil this role, from alchemilla to ammi, *Salvia viridis* blue-flowered to scented-leaf pelargoniums, such as 'Attar of Roses'. It's good to choose a different colour and form to the primary foliage, using it to fill up any prominent gaps and unevenness.

Upper storey: This third foliage plant should have an interesting architectural shape. The idea of this is to break up any over-neatness, introducing plenty of ups and downs to mess the whole thing up a bit and make the arrangement as three-dimensional as you can. The elegant curving flowers of honeywort, an interesting grass such as *Panicum capillare* 'Sparkling Fountain', or a seedpod, such as a green honesty pod, are excellent for this, as is a green vertical spike, such as bells of Ireland.

Vine: This is not essential, but adding a vine to your foliage creates a relaxed shape as it cascades down the arrangement. I love things like perennial sweet peas, cobaea and cardiospermum (see page 248).

A hand-tied bunch of July-harvested hardy annuals following the 'Bride, Bridesmaid and Gatecrasher' rule.

Flowers

You'll need about seven stems of the bride, nine or eleven stems of the bridesmaid and five or seven stems of the gatecrasher. When arranging, add the flower stems at an angle, not straight down, starting with the bride.

Bride: This is what I call the dominant flower of the bunch. This is the flower that you fall in love with in the garden or flower shop, the one you've got to have. It forms the centre, the pivot of the arrangement.

Bridesmaid: This is the same colour or very similar to the bride, but a smaller, less glamorous flower that is not as dominant. It should not compete with, but back up the bride.

The gatecrasher: This is the contrasting flower that brings the whole thing to life. This needs to be a colour that clashes with the two other flowers. My arrangements are not about calm, recessive harmony, but about being centre stage – and the gatecrasher is how you achieve this. I think of this like a squeeze of lemon with smoked salmon, the all-essential contrast that makes the whole thing work.

August

Colour still has a huge presence throughout August. Much of the garden is packed with half-hardy annuals such as cosmos, cleomes and zinnias, as well as a crowd of tender perennials – not just dahlias, but osteospermums, argyranthemums, petunias, verbenas and pelargoniums in our plethora of pots. These are all pumping out the flowers, yet August can still be a bit disheartening in the garden.

I remember the days when it was just me (working part-time) and one other gardener looking after the intensive one and a half acres (or 1.62 acres to be precise), and it was impossible to keep up in late summer. Flowers were still going, with more to come, but we struggled to stave off the garden's tendency to sag. I work with a bigger team now, but in August, it still takes copious amounts of time to keep the place looking fresh, flowery and vibrant.

Most of our roses are still flowering lightly, but certainly from the middle of the month, the flowers become fewer. Even the mid-season flowering perennials – agapanthus, phlox, heleniums, crocosmias and stipa – are fading, or at least beginning to develop seedheads and look autumnal. And we're not ready for that yet!

Thank goodness there are enough plants which come into flower now and carry on with brilliance until the late autumn. We make sure to have plenty of those at Perch Hill. We have our carnival of dahlias providing a procession of colour (more on those in September), and salvias by the barrow-load. They are invaluable for keeping the garden looking cheerful late in the year – and luckily there are plenty of other things coming into their own now, or still excelling.

Mexican fleabane (*Erigeron karvinskianus*) is a true stalwart, pretty much all year. Apart from in the Oast Garden and Dutch Yard (where the colours are mainly bright or rich), we encourage it to grow almost everywhere except the borders, and let it self-seed to soften the stone and brick paths, steps and walls with carpets and curtains of white daisies. It's one of those remarkable plants that just keeps on flowering.

We also have groves of hydrangeas. These are usually associated with autumn, but it's now that varieties like *Hydrangea paniculata* 'Limelight', *H. paniculata* 'Wim's Red' and *H. arborescens* 'Incrediball' provide that radiant, soft-green colouring. We have a walkway lined with them in a dark part of the Dutch Yard, with several 'Incrediball' planted just outside our east-facing kitchen window. When I'm standing at the sink, at a right-angle to that wall, I find myself doing a double take, imagining there's someone at the window. Their flower globes stand up pretty straight (unlike its often-droopy parent, 'Annabelle') and are a very similar size to a human head.

Then there's the great band of half-hardy annuals – planted out in May (see page 157), they are at their peak. Cosmos are our undisputed half-hardy-annual August queens. We concentrate these in the cutting gardens, but they give us an invaluable colour-sprinkle all round. Verbenas of different colours, shapes and sizes are everywhere, and we have tithonia and antirrhinums scattered liberally.

Below An avenue of hydrangea on the north side of our house, with *H. paniculata* 'Limelight' and *H. arborescens* 'Incrediball' in the shade.
Opposite A bunch of mixed cosmos 'Purity', 'Sea Shells' and 'Fizzy Rose Picotee'.

Our pots in the Dutch Yard full of smaller dahlias (including 'Schipper's Bronze', 'City of Alkmaar' and 'Dahlegria Tricolore'), as well as chocolate cosmos and Spanish flag on the silver birch teepees beyond.

Our pots also take centre stage. Ten years ago, we had twenty pots grouped in one or two key places, but slowly over the years we've added more and more. They're everywhere now, three hundred in all, including the small ones. We have a line of large pots in the rose garden, a big collection on the stone slabs outside the barn and on the oak table in the middle of the lawn. We have pots along the edge of the lawn, tall Danish pots in the Oast Garden and in the Farmhouse Garden, more at the entrance to the Perennial Cutting Garden and down under its arches, and around the brick plinth of the Chelsea shed. On and on they go, nowhere more so than in the Dutch Yard outside our back door.

Pots in August, if we've kept them in good nick, are looking their summer best. They have a high point in April with tulips, and another now. The plants in them have grown into each other, reached their full height and are tumbling and trailing down the full length of the pot. And with towers of silver birch teepees rocketing out of the top of our larger pots, clad in gorgeous colour from our ever-increasing range of tender perennial climbers, we've got an August parade of pots.

Opposite A favourite pot on our doorstep surrounded by carpets of *Erigeron karvinskianus*. It contains *Osteospermum* '3D Banana Shake', gazania 'Blackberry Ripple' and the smaller-leaved helichrysum 'Silver Mist'.
Below *Erigeron karvinskianus* is not only good in the cracks to soften steps, terraces and new paths, but is one of the prettiest, longest-flowering living table centres too. We grow loads.

Tender perennial climbers

When I walk out of our house into the Dutch Yard and on to the Oast Garden in August, I want the place to feel as if it's brimming over, intensely colourful, stimulating, frothy and plant-rich, with pots gathered at entrances and along path edges adding extra oomph to the borders.

With exactly this in mind, more than twenty years ago, I laid out the Oast Garden and wrote my book, *The Bold and Brilliant Garden*. I wanted to achieve a sense that the plants were dominating us, rather than the other way around – and I hope this feeling remains.

From August it can be a challenge, but with plenty of tender perennial climbers up our sleeve (mostly grown as annuals, because they won't survive the winter), we can achieve it. There's no group of plants more important for jungle-like fullness at this time of year; they have huge presence. Trained up the silver birch or hazel twiggy teepees that stand on top of our large Danish pots and long toms, they create towers three-metres tall – great pyramids and curtains of brilliant colour and form.

I love a climber. We rely on their ability to create a room-like enclosure, and there's something even more marvellous about that when it's achieved with an annual. They inspire in me some deep level of amazement at what a seed can do in less than a year. You can have a French bean reach four metres and a cup and saucer vine poking in at the attic window by September – they go from next to nothing to a plant that towers above you in a matter of months. This is all very well for a clematis, whose roots remains in the ground to grow and strengthen from year to year, but for a tiny seed to do this at such a lick seems miraculous. It's perfect for us in August, exactly fitting with what we need in the Dutch Yard and Oast Garden.

Every one of these plants flowers for ages, without much need for deadheading. The cup and saucer vine (*Cobaea scandens*) is the most remarkable of all. One winter, when we had no hard frost, it not only survived, but flowered lightly right through until spring, when off it went again to give us a fully clad flowery curtain.

This has become a classic mix here: *Rhodochiton atrosanguineus* 'Purple Bells' with *Thunbergia alata* 'African Sunset' on a silver birch teepee.

Cobaea scandens with *Asarina scandens* 'Mystic Violet'. Both start to flower prolifically in mid-summer and continue until at least December (see picture on p 340).

The Chilean glory flower (*Eccremocarpus scaber*) does much the same, but in a powerful mix of flame-like colours. I've picked a bunch of cobaea and eccremocarpus growing up a sheltered wall on Christmas Day. Climbing snapdragons (*Asarina scandens*) are also a highlight. They survive with us in a sheltered spot from one year to the next, so retain incredible heights and flower every summer.

Then comes purple bell vine (*Rhodochiton atrosanguineus*) and the black-eyed Susan vine (*Thunbergia alata*), a favourite being 'African Sunset'. Together, they make a truly beautiful pairing. This would be my desert island duo, and I often grow them on the same frame – the centre of 'African Sunset' flowers match the black dangling centres of the bell vine. Underplanted with the mahogany velvet lady's slipper (*Calceolaria* 'Kentish Hero'), which is now perennial in our protected pots and borders, it is one of my favourite plant combinations of the whole year.

The most delicate of the lot, but wholly worth it, is love-in-a-puff (*Cardiospermum halicacabum*). This is almost a weed in a friend's garden in New York State, where summers are hot and sunny; here it's not so prolific. We always try to find room to grow this in the greenhouse (where it now self-sows), as there's nothing better for autumn and Christmas wreaths (see page 372). But I still have to sow and grow this every year outside. I'd miss those bright apple-green seedcases trained over our chestnut paling fence and trailing out of the side of pots like mini-lightbulbs. They keep their colour until November.

The towering teepees are like upside-down witch's broomsticks, which we make every spring with the twiggy tops of silver birch branches (see pages 88 and 124). These have so much more character and beauty than a smooth-skinned, dead-straight bamboo. As with a local stone or brick wall, they feel part of the place, a connection between the cultivated and the wild, and they're a hugely important part of the garden here through late summer and autumn.

With every one of these tender perennial climbers, you get double or triple flower value compared with hardy perennial climbing roses, clematis or honeysuckles. We'd lose a lot of colour – and form – if we didn't grow plenty of them, and we've started to try each one not just as a climber, but as a trailer, too.

Best of the tender perennial climbers

We use an ever-increasing range of these long-flowering climbers. Between them, they clad the arches in the Perennial Cutting Garden, alternate with sweet peas on teepees down the main path in the Annual Cutting Garden, clamber all over the pot frames in the Dutch Yard and cascade down the sides of our water troughs, window boxes and huge Danish pots all round the garden.

1&2 *Asarina scandens* (climbing snapdragon)
I grow these in three colours: white (*A. erubescens* 'Bridal Bouquet'), soft mauve (*A. scandens* 'Mystic Rose') and deep purple (*A. scandens* 'Mystic Violet'), and I've fallen for them all. We mix 'Bridal Bouquet' with the white-flowered cup and saucer vine (*Cobaea scandens* f. *alba*) in the Perennial Cutting Garden on metal arches, and have 'Mystic Violet', now seemingly perennial, climbing up a sheltered east-facing wall so that it reaches up to Adam's first-floor workroom window. It can flower there until after Christmas and is back in flower again in May. I can't think of a climber that gives more.

3 *Cardiospermum halicacabum* (love-in-a-puff)
I love one of the common names of this marvellous climber, 'love-in-a-puff'. It has a jet-black trio of seeds inside every green football-like seedpod, each one stamped with a perfect white heart on its side. Its Latin name also means 'heart seed'. We love it.

4&5 *Cobaea scandens* (cup and saucer vine)
The flowers of this vine look exactly like their name: a small cup sitting on its saucer. It's too strong and triffid-like for a birch teepee – it puts on so much growth a gust of wind makes the whole thing take off like a sail. Instead, this Mexican native clads our metal arches in the Perennial Cutting Garden where I pick it until Christmas and beyond.

It's one we've been trialling as a trailer, rather than just a climber. The seedlings were planted in 45cm (18in) deep water troughs on the first-floor balcony in the Oast Garden, and they ended up tumbling, unsupported, over the side. A friend claimed she'd had them in a window box in London and had to give them a haircut when her neighbours three floors down complained it was obscuring their view. We copied her idea, and ours now hangs over the bathroom window. It beats net curtains any day! (For planting cobaea, see page 54.)

6 *Eccremocarpus scaber* (Chilean glory flower)
Perennial in a sheltered spot, this climber looks great against red brick or cascading over a wooden fence. Its mix of coral and orange tubular flowers is super-long-flowering and it's very easy to grow.

7 *Ipomoea lobata* (Spanish flag)
Pretty, delicate, Chinese pagoda-like flowers that cascade all the way down the quick-growing vines look so sweet picked and arranged in a small jug, their stems first seared in boiling water. This is one of the fastest of the annual climbers to put on a show.

8 *Ipomoea purpurea* 'Grandpa Otts' (morning glory)
It's tempting to grow the classic blue morning glory (*I. tricolor* 'Heavenly Blue'), but in our trials here, it never does well. A few autumns ago, I came across an entire acre of its brilliant blue flowers and dark, snaky green vines on the island of Ischia in southern Italy. It was a Rousseau-esque sight that any gardener's heart would have leapt at, so I tried planting it yet again. It hardly flowered and the few flowers

we had stayed open for less than a morning. 'Grandpa Otts', on the other hand, with its deep purple flowers that stay open until mid-afternoon, is the one to grow. It's been bred to be more compatible with the UK climate.

9 *Rhodochiton atrosanguineus* (purple bell vine)
Another tender perennial native to Mexico, this grows quickly enough from seed to be hugely worth growing as an annual (we sow this in February). It has heart-shaped leaves with stems that twirl around whatever you lean them against, the new growth green, inked with purple. As well as writhing upwards, newly formed shoots hang free like purple ribbons around a May pole, and so we are trialling this as a trailer for our larger pots and window boxes.

The flower is in two parts: the tubular centre (the true flower) hangs as the almost-black clapper in a blushing-purple bell (which is the calyx). *Rhodochiton* is from the Greek, meaning 'rose tunic', and even when the dark central tubes have dropped, that rose tunic goes on elegantly for weeks.

10 *Thunbergia alata* 'African Sunset' (black-eyed Susan vine)
A sophisticated version of the good-old black-eyed Susan, 'African Sunset' has beautiful apricot and peachy flowers. It opens a flame orange and then softens as the flowers develop, giving a lovely mix on one plant. We did a trial of this family last year and also loved the deeper, richer mahogany 'Tangerine Slice A-Peel'.

Cosmos

It's easy to feel snooty about cosmos, but don't. They are the best all-rounders for high-summer picking and last-minute border filling. Our August garden would be a lesser place without them. Just like tulips, roses and dahlias, cosmos feel like bread-and-butter to the colour- and flower-loving gardener. I personally owe cosmos a lot. They are the foundation stone of much of what I've done in the last three decades.

I trained as a doctor and I loved my day work on the wards, but hated working through the night away from home. I've never been a night person. After I had my first child, my hatred of night duty was accentuated. I became desperate to think of ways I could be at home. As a lover of cut flowers, but with no overall plan, I started growing scrappy lines of annuals to pick for the house – things like English marigolds, cosmos and poppies, seeds I'd got from the local garden centre.

At that point, in 1995, in my early thirties, I had hardly sown a packet of seed, but even with these easy chuck-in annuals, it was quickly clear to me that some plants survived on near neglect, flowering for three months or more (cosmos), while others (opium poppies), lasted barely three weeks. Admittedly, neither got any staking, they had minimal water and only sporadic picking – I had no time and almost no knowledge. But they responded differently to this lack of TLC. Once inside, the poppies also dropped their petals instantly. I liked them, but they seemed a waste of time as a cut flower.

The following spring I ordered a seed catalogue. I knew wildflowers and the cultivated plants from my parents' garden, but those were predominantly perennial. This annual seed-grown stuff was a whole new world to me. So I looked out for the scissors symbol (the marker that they were good cut flowers), and chose a few varieties.

I selected *Bupleurum rotundifolium*, for its lovely bright green foliage; *Euphorbia oblongata*, as I knew a few perennial euphorbias from my parents' garden; cornflowers in blue and

A huge vase of cosmos 'Xsenia', 'Purity', 'Candy Stripe' and 'Sea Shells' in our kitchen.

black, as they seemed jolly; and a plant called *Ammi majus*, which looked like cow parsley, but I'd never heard of it before. I also added a marigold (this time in a more sophisticated mix of orange and crimson) called 'Indian Prince', together with a couple of cosmos ('Purity' and 'Dazzler') – these were chosen on the grounds that they'd done well the previous year.

I chose a spot between some blackthorn bushes in the not-yet-existent Farmhouse Garden. Clearing away the grass and digging it over a bit, I roughly prepared a mini cut-flower patch. To get a bit more of a system going than in the previous year, I marked out a series of one-metre square blocks with bamboo canes.

In May (as it advised on the back of the packet) I sowed each pack into their own square. My elder sister, Anna, told me it was best to sow the seed in lines, rather than broadcast, so I'd be able to distinguish sown plants from weeds – and that I'd need to keep on top of the latter.

In a hit-and-miss way to begin with, using my instinct rather than a learned method, I tried to record how long each variety looked good for. Also, because I loved picking flowers,

We plant this combination a lot: *Ammi majus* and cosmos 'Purity'.

I noted down every handful I brought into the house. And I tried to record the date they were first cut and conditioned and put into water, and then the date they'd gone over and I'd chucked them in the compost bin.

Gradually this evolved into a sort of marking system, showing how long each plant took to grow from seed to flower, then how long they flowered for. As I got more experience sowing and growing, I also rated how easy they were to grow. And taking all these things into account, I felt I was getting an overall feel for square-metre productivity.

I also knew from picking hellebores as a child that if I seared the stem ends in boiling water, most would cut pretty well. Without this boiling water plunge, they flopped straight away, so I started searing every stem end of all the summer plants and recorded their improved vase life (see page 231).

Between productivity and vase life, and by awarding an extra point for any remarkable characteristic such as lovely perfume, I ended up with an overall mark per plant. It'll be no surprise to anyone who has grown it that cosmos 'Purity' came out on top, with 'Dazzler' next. They were the two clear champions.

To get the most out of the cosmos, as well as shave the time it takes for them to go from seed to flower, I no longer direct sow them in the garden. Directly sown in April, they won't flower until August, but if sown under cover in late March or early April, and transferred to the garden in early May, cosmos will flower by July or even late June. They will usually go on to flower for at least sixteen weeks.

In my first ever cut-flower trial, cosmos gave me three milk pails worth of flowers for most weeks, adding up to just under fifty buckets per square metre. The foliage is elegant, light and feathery, so you don't even need extra greenery to mix in – they look good simply arranged in a big vase on their own.

The flower's egg-yolk-yellow centre is stuffed full of nectar and pollen, so as soon as the sun comes out, they're heaving with butterflies and bees. It all makes cosmos a wonderful plant for the garden and house.

It was this first cosmos triumph that really kick-started my professional gardening life. Concentrating on the plants that had done best in my garden, I launched a catalogue after three years of simple trials – and cosmos were the first to go in.

Drawn by an artist friend, Flora McDonnell, the first ever catalogue had a monochrome illustrated flower on the cover that was stapled onto one A4 sheet of paper with a list of seeds. I printed just one hundred of these, and I sat at our kitchen table and hand-painted every one of those black-and-white drawings.

That first catalogue had 30 varieties in total. All the seeds were individually counted or weighed by me, family and friends. Now, 25 years later, we have nearly 30 varieties of cosmos alone – big ones, bright ones, pure-looking ones, great lasters, great flowerers, an entire universe of cosmos. It's all down to a few square metres of trial gardening, and I feel proud.

Below The newly bred cosmos 'Xsenia' with the self-seeding fishing rod-like flowers of *Persicaria orientalis*.
Opposite Cosmos 'Double Click Cranberries' and 'Candy Stripe'.

Best of the cosmos

Whenever I talk to any beginner gardener who fancies sowing something from seed – young or old – I encourage them to start with cosmos. Pretty much every variety I've tried has been fast to germinate, easy to grow and gives so much for so very little.

1 *Cosmos bipinnatus* 'Dazzler'
'Dazzler' is a good name for this one; a brilliant magenta pink with big saucer flowers and golden hearts. I have to confess, my love of 'Dazzler' has transferred to the richer coloured 'Rubenza', but I still feel an old affection for this classic and grow it often.

2 *C.* Double Click series
Unlike many double flowers, the Double Click varieties contain pollen and nectar, as not all of their nectaries are bred to be extra petals. This gives crumpled silk handkerchief flowers for us and food for pollinators. There's the rich 'Click Cranberries' (pictured), and I like the white 'Psyche' or 'Fizzy White'.

3 *C.* 'Purity'
Needs no introduction. No garden should be without this plant: it's still the strongest, longest-flowering and most vigorous cosmos when compared to new ones in a trial. It self-sows in a sheltered spot here.

4 *C.* 'Rubenza'
Opens very dark, like crimson velvet, and then brightens as the flowers develop, becoming similar to 'Dazzler'. I love that evolution and the mix of colours you get in a patch of a few plants.

5 *C.* 'Sonata White'
This is like 'Purity', but less than half the size, so ideal for filling window boxes and containers from summer into autumn. It's pretty on its own or pierced through with the lovely spires of *Gladiolus murielae*.

6 *C.* 'Xanthos'
This is quite a newly bred form in primrose yellow that's rather in fashion. We have found it gets rust more readily than older forms, but are trialling it again and giving it a regular dousing of chive tea (see page 193).

7 *C.* 'Xsenia'
Another fashionable newcomer that's in a softer colour than most of the non-white cosmos. Flowers open pink and develop a pretty apricot tinge to the centre of each petal that makes them look almost translucent.

August dawn in the Rose and Herb Garden with salvias 'Amistad' and 'Jezebel' flowering away.

Practical
August

August tends to be about a lot of little jobs, with some regular chores such as taking cuttings, deadheading, feeding pots and watering, rather than one enormous task we need to undertake. It's really worth looking after our pots to keep them going for a couple more months, as they fill the place with life and colour.

Late-flowering perennials such as phlox and Japanese anemones can struggle a bit in the heat, as do floriferous shrubs such as our hydrangeas – they literally start to sag. We leave a hose at their roots once a week, with water just gently seeping, for ten to fifteen minutes so that it penetrates deeply and helps sustain these moisture-loving plants through into autumn.

As growth slows, it's also time to clip our evergreen box balls and yew. Box blight is on the increase, and since this is a fungus, we try to make sure there is good air circulation on the border-side of our plants, as well as the more open path-side – this is key in blight prevention. The fungus loves a tightly clipped box, so we leave our balls slightly looser and carefully remove some of the inner branches to increase the air flow.

We cut our quick-growing hawthorn hedges twice a year – in winter and in August. We give them an undulating top, as I like their rhythm with the surrounding countryside, and now is the time to shape them. First we check that any bird nests are empty, then we cut them back hard to prevent them growing too tall and creating too much shade for the plants at their base.

We also start to plant out early-flowering biennials (honesty and wallflowers) that were sown under cover in May and June. We find if we plant these in their final flowering position during August, we're guaranteed whopper plants before the late-autumn cold sets in. Their roots will develop quickly at this time of year, as the soil is warm and increasingly moist at night with the start of the proper dews. We want to use this naturally perfect growing environment to our advantage for as long as we can.

We mow our wildflower meadow anytime from the end of July. By doing that, we can be sure that even the orchids and the yellow rattle have ripened and dropped their seed. We are protective and proud of two clumps of pyramidal orchids in our grass, but they're dying back by August, so it's safe to chop. We cut (with a strimmer) and then leave what we've cut for a couple of days to dry and drop any residual seed, then we rake up the hay and remove it. The key to the success of a wildflower patch is to continue to decrease the fertility of the soil, so don't cut and mulch with grass clippings. To prevent the newly shaved look, we mow on the highest blade setting with our mower.

And there's one final thing on our check list: if we have hyacinth bulbs stored from last year, we put them in a paper bag at the bottom of the fridge. We leave them there for a month, pre-chilling before planting, and then plant them in bulb bowls in September (see page 298). We store the bowls in the cold and dark, and these will be in flower by Christmas. Doing this means we avoid the need to buy more expensive forced bulbs that have had this chilling process done

Watering and feeding, particularly shallow pots, is quite a chore in August, but it's worth the effort to maintain them so they can continue to put on a show.

for you. I find if I look for hyacinths to buy in August, they're surprisingly hard to come by. The ideal thing is to preserve our own from one year to the next.

Taking Care of Pots

We really go for it on pot care. Without it, our August and autumn garden would really be diminished – but boy-oh-boy, do pots need watering, feeding and regular deadheading.

Watering

We find it's worthwhile lining our large terracotta pots with empty plastic compost bags as it decreases evaporation from the pot sides.

Small table pots dry out so quickly when it's hot, we sit them in saucers so they can continue to absorb the water that's initially drained away. If it's really dry, we plunge the bottom of the pots into a bucket of water until the bubbles stop rising from the compost.

With our larger pots, Josie, our head gardener, uses this triple watering technique which has to be done daily if it's hot and dry. Otherwise, on duller or cooler days, we stick a finger in the compost to see if it's damp and water according to what we find.

❀ Use a hose pipe, not a watering can, as it's more efficient for the volume of water required. Be aware, if you leave your hose unwound, particularly if it's black, the first

water that comes out after a hot day can be so hot, it will damage the plants, so let the first minute or so of water just run away before you start watering.

❀ Step one in the three stage system is to water every pot in the group thoroughly for a minute or so, giving them a real drenching. If the container compost has dried out, water quickly runs out of the drainage holes and it's easy to think this means the container has had enough, but that's usually not the case. When dehydrated, the compost can't hold on to moisture. If it's rained in the last day or two, this first stage may not be necessary. Josie also recommends pressing the compost against the edges of the pot if it has shrunk away, to help it absorb the water.

❀ The second stage is to water the group of pots again, but not for quite so long this time. Water them for 30 seconds.

❀ And then finally, water them a third time, so everything gets three solid waterings. By then, the whole plant root is well-hydrated with water right down to the micro-roots.

Feeding

❀ Compost usually contains enough nutrients to feed plants well for 4-6 weeks, so May-planted pots need to be fed starting in July.

❀ By August and September, every pot is fed once a week. We use liquid seaweed feed. For the first six weeks or so, we do so at the strength recommended on the bottle.

❀ By September, we move to a double

Deadheading dahlia 'Schipper's Bronze'.
Flowers going over, as opposed to new buds,
are pointed not round.

concentration. For window boxes (where there's minimal compost below the roots), by the autumn, we use triple-strength liquid seaweed feed.

❧ If the weather is very dry, we spray with sulphur-rich chive tea (see page 193). This helps protect against mildew, which can become an increasing problem in a dry year with plants such as container-grown dahlias.

Deadheading

Regularly picking or deadheading half-hardy annuals, dahlias and tender perennials such as osteospermums, arctotis, argyranthemums, gazanias, verbenas and pelargoniums is important to ensure they go on flowering and looking beautiful right into autumn.

❧ If I'm in a hurry, I sometimes just pull the larger spent heads of our osteospermums, arctotis, argyranthemums and gazanias as I walk past, but it's better practice to cut back the ageing flower to a leaf or to an already formed axillary bud.

❧ With dahlias in particular, people are sometimes anxious that they won't be

able to tell the difference between a spent flowerhead and a bud, and it's true that with certain varieties, such as 'Schipper's Bronze', it can be difficult to distinguish. A deadhead has a pointed shape and the bud is round.

✤ We try to do this once a week or every ten days in August, but it's a good idea to always have snips in your pocket and deadhead whenever you have a spare minute and pass a plant that is in need of a bit of tidying. An ideal moment to do this is as you're watering.

Taking Cuttings

People are often surprised when they walk into our polytunnel in August that it's chock-a-block with pots of cuttings. Many gardeners take their cuttings in spring, but we find it's now, when many of the tender perennials are growing at full tilt, that they root the fastest.

Many of the plants we grow here – the pelargoniums, osteospermums, argyranthemums, arctotis, nemesias and diascia – come from South Africa, and it's in August and September that they start their main growth spurt, coinciding with their native spring. So propagating now (rather than in our spring) makes good sense.

We also propagate plenty of rosemary and lavender in August, to give us plants for next spring, in case we lose some (as we often do). The rosemary and most of our lavender (such as the English lavender) has finished flowering and needs cutting back. We cut all the flowers off and prune above new shoots, not cutting back into old wood. We end up with a compact, rounded plant that will put on some new growth before winter. We can then do another prune in early spring if necessary, leaving the old wood.

We take cuttings in late spring and early summer. These are ideally at 5-6cm (2in) long. There are instructions on the following pages for general cuttings, but the only thing to bear in mind with lavender is to make the cutting by pulling a side shoot from a main stem with a small heel of bark attached to it, rather than cutting from the main stem with scissors.

We take salvia cuttings a bit later, in September or October, along with plectranthus, calceolaria, cobaea, fuchsias and heliotropes. That's as a result of trialling cuttings taken in summer and in autumn – later gives us faster rooting with this selection of plants.

Taking cuttings from tender perennials
(such as this *Heliotropium arborescens*
'Reva') is one of our main jobs at this
time of year.

Tender perennial cuttings

❀ Take a short piece of stem from the
main plant.

❀ Trim to just below a leaf joint, so the cutting
is 5-6 cm (2in) long. Short cuttings root
better than long ones. Just below a leaf node
is where there is the highest concentration
of natural rooting hormone. We don't use
rooting hormone powder as we find all our
cuttings root consistently and quickly
without it.

❀ Strip off all leaves except the top pair. If
the top pair of leaves is large, cut these in half
across-ways. This seems brutal, but they will
stay alive and continue to photosynthesize,
and with half the surface area they won't
place as much demand on the stem to draw
up water to support them. This makes the
cutting more likely to root rather than flop.

❀ Remove the stem tip. It's at the top that the
growth hormone concentrates, so by pinching
it out, there's nowhere for it to go but down
in order to encourage root formation.

❀ Fill a series of small pots with a mix of
1 part grit to 2 parts compost.

❀ Insert the cuttings round the edge of the
pot, spaced about 4-5cm (2in) apart (see the
illustration on page 267). By placing them
around the edge, not in the centre, you
encourage quicker root formation, as
the new roots quickly hit the side of the
pot, break, and then branch into more
lateral rootlets.

❀ Water the compost well.

❀ Place the pot undercover, on a heated base
(see page 79). This speeds up the rooting
process. Black plastic absorbs the heat and
promotes rapid root formation.

❀ With rosemary and lavender cuttings, put
them in a cold frame or a cool, but well-lit
spot and keep the pots well-watered.

❀ Put the pots on a bed of grit (or capillary
matting) for the next couple of months and
water only when the compost is dry when
you poke a finger down 5cm (2in) below the
compost surface.

❀ Cuttings often root in three or four weeks.
Check for roots showing in the drainage
holes, and once formed, pot each cutting on
individually.

❀ If any of the cuttings show the slightest
sign of botrytis (browning of the cutting
or visible mould), take it out as this will
spread quickly.

❀ Store them under cover through the winter,
for planting out in spring.

September

We have hawthorn everywhere at Perch Hill. When we first arrived, I suggested planting clipped yew down the drive, but Adam (rightly) said it was too grand for a farm garden. Instead, we went for common hawthorn (*Crataegus monogyna*).

We put in twenty trees, alternating them down the sides of the track. Blasted by wind from the west, they've only grown to around four metres in twenty-five years. They are anything but smart, with many of them now leaning from the force of the gales, but draped in lichen and dense with haws, I love them. They form an important part of the September garden – red, but not too red.

More recently, we added the broad-leaved cockspur thorn (*Crataegus persimilis* 'Prunifolia') to our Dutch Yard, bang outside our first-floor bathroom window. It's one of the great autumn trees, with beautiful and long-lasting leaf colour and crab apple-sized haws until at least November. Even when these fall to carpet the soil beneath, or drop into the old sink that's the dog water trough, they keep their colour and make a great sight. Frozen after a cold night, they're like rubies in silver.

During our first couple of years here, we planted half a mile of hawthorn hedges. Perch Hill is well-named, the house and garden face south, tucked into the side of a hill. It's very exposed and our soil is impenetrable Wealden clay. If you dig a hole, you can make pots from the stuff that comes out, and in many places we have only an inch or two of topsoil, black and rich, sitting on the yellow clay. To make a garden out of the field and farmyard, I had to improve the soil structure massively with great quantities of organic matter and grit, and enclose every area with hawthorn. This does well on clay and is the quickest, most affordable way to provide shelter.

Hawthorn's a pig to work up against (you inevitably get pricked when weeding or planting nearby and there are thorns from hedge clippings buried in the soil all round), but the whips grew into decent-sized hedges within three years. We've now clipped the hedge tops to create undulating curves, relaxed and imprecise, rising up over all the gates to form arches. It makes a softer and more interesting backdrop than a hedge with a short back and sides.

Contained and protected, the garden is brimming with flowers and colour in September. There's no fall-off yet. In fact, I'd like to claim that whether you visited in April or early October, you'd find the same concentration of colour. We always keep the garden open right to the end of September; the National Garden Scheme struggles to find gardens that volunteer this late in the year, but it's one of Perch Hill's best moments.

The dahlia colours start to soften and fade a little in October, but in September they still carry their full punch and we have them crammed into flowerbeds and containers, more and more added each year from our trials.

Below Spindleberry and hawthorn, are both autumn-fruiting plants that are important for feeding our birds. Opposite Our original dahlia trial garden with 'Profundo', 'Purple Haze', 'Burlesca', 'Black Jack', 'Rancho' and 'Karma Naomi' below a line of common hawthorn.

The Oast Garden at its peak
of floweriness and scale, with
Rheum palmatum, dahlia 'Bishop's
Children' (in pots), *Rosa moyesii*
and *Helianthus* 'Lemon Queen'.
There's also *Petunia × atkinsiana*
'Tidal Wave Purple' and the hanging
curtains of *Cobaea scandens*, which
reach almost ground level by the end
of the year.

And on the cut-flower front, we have guilt-free armfuls to pick. Dahlias are joined in some places by their mini counterpart, the zinnia. By sowing the right varieties in late May (any in the Benary's or Queeny series), we have them at their flowering peak now.

As well as all the different colours of cosmos, which are still completely happy, and their attendant group of half-hardy annuals (nicotiana, cleomes, sunflowers and verbenas), we also have lots of tall gladioli. Anything without a vertical accent – whether it's a landscape, painting or garden – looks empty, as if it's missing something. Vertical spires create a visual structure on which a sense of balance and coherence relies. That's why I plant dahlias with giant teepees of climbers and groves of gladioli. To paraphrase Vita Sackville-West, who was referring to her rose/delphinium/foxglove combination: they are the all-important minarets to our dahlia domes.

We choose coloured glads to harmonise or contrast with the dahlias. They make each other seem at home and we've found that if they're generously mulched and winter-protected (in at least 15cm/6in of compost), both gladioli and dahlias emerge happy and hearty, like true perennials, each spring.

Opposite Picking rich and brilliant-coloured gladioli for a large party vase.
Below Our Rajasthan dahlia collection, presided over by the crazily vast and marvellous dahlia 'Emory Paul'.

Dahlias

'Rip City' was the first dahlia I grew; I bought it from the nursery at Monet's garden in Giverny in 1999. Dahlias were out of fashion then, but the sight of 'Rip City' growing in the 'paint box' beds in late September instantly converted me. I loved this one's rich, plush texture and the dark colouring of its flowers – and the scale of the flowerheads was impressive, the size of a small plate.

I bought three plants, stashed them away in a cold frame until the following spring and then planted them with the similarly textured Mexican sunflower (*Tithonia rotundifolia*) and some matching chocolate cosmos (*C. atrosanguineus*). The dahlias romped, growing to nearly my height, with just those three original plants of 'Rip City' giving bunch after bunch of flowers every week for twenty weeks or more. That was it for me. I've been a big dahlia fan ever since.

There are a few ugly dahlias, with too much going on (I'll name no names, in case I later change my mind!), and it's true that they can get munched by slugs (see page 292), and I know that some people don't like the smell... but you can't let these minor things put you off.

Our newly mild winters mean that dahlias can be treated much like a perennial; in all but the coldest, highest and wettest spots, they can be left in the garden through the winter in the UK – and there is then no more generous, lower-maintenance flowering plant.

Dahlias are all prolifically cut-and-come-again. Whether you're liveheading or deadheading, by removing the leader of any stem, you are encouraging axillary buds to develop below and more flowers to form. The lower you pick in the plant, the longer the delay until the next lot. I regularly pick three foot stems for huge vases at home and that means I won't get another flower on that part of the plant for two or three weeks, but it's always worth it. I just need more plants.

I harvest in the cool of the evening or early in the morning when the flowers are fully open, but the petals at the back

Some of the best soft and warm dahlias (see p286) including 'Labyrinth', 'Penhill Dark Monarch' 'Molly Raven', 'Perch Hill', 'Café au Lait Royal' and 'Preference'.

still look fresh. On average, dahlias picked at this time of day should have a vase life of four days. There are a few that do better: 'Perch Hill', 'Dark Butterfly', 'Karma Choc', 'Zundert Mystery Fox', 'Raffles', 'Molly Raven' and 'Daisy Duke' all last nearly a week in water.

Most of those in the Single-flowered group are generally not so long-lasting, looking good cut for only a couple of days, and those in the Anemone-flowered group, such as 'Blue Bayou' and 'Totally Tangerine', don't make good cut flowers either. They're best left in the garden, but look magnificent there.

Since planting 'Rip City' twenty years ago, we now have more than a hundred different varieties. We have dahlias in the herbaceous borders, dahlias in pots, dahlias in the cutting garden and on the veg and salad bank, and I even decorate our salads with a jazz of dahlia petals (which are edible). We've run a dahlia trial every year for more than a decade and in 2018 we created a new garden almost completely devoted to dahlias. In 2020, we added two more vast dahlia beds, both fifteen metres wide. You can't move for dahlias here in the autumn.

The Annual Cutting Garden path lined with dahlia 'Happy Halloween' and 'Blue Bayou', leading up to our Chelsea shed (from our RHS Chelsea Flower Show 'Colour Cutting Garden' in 2017).

Supremely easy to grow and spectacular in a vase, dahlias are, without doubt, my favourite plant and we have now bred several of our own. As with tulips (see page 99), the process of selecting our own named dahlia hybrids starts with an early autumn visit to various dahlia breeders, usually in Holland, where we walk the fields to look at the new hybrids that are being bred.

We choose a few stand-out beauties when they are just seedlings. The following spring, the breeder takes cuttings from that plant and grows it on in a small group. At that stage, the resultant fields of dahlias are an incredible sight, with a hundred or more new or almost-new varieties, just a row or two of each, growing cheek by jowl. I sometimes select varieties at this second year breeding stage, too.

We bring the first or second year dahlias we've selected back to Perch Hill to trial them. In the following season, we decide if we still like and rate them, whether they have any stand-out characteristics, whether they're light or prolific flowerers, and what their vase life is like. We assess whether they have good mildew resistance (a problem in dry summers) and general vigour and health. Once we've decided we want to back a dahlia, its tubers are divided by the breeder and more cuttings taken, and on and on until we can fill a field.

Then we name and release it, proud parents of our dahlia offspring. We now have a dahlia named after Perch Hill and another after me. Another has been named after my long-standing business partner, Lou Farman, and one after my youngest daughter, Molly. Recently we've named one after our head gardener, Josie. And there'll be more!

Best of the dahlias

There are several groups in the dahlia clan. My favourites are Cactus (which are spiky), Single-flowered (as they sound) and Anemone-flowered (which look like they have a flower within a flower). When I'm designing combinations with dahlias, I don't find these groups as useful as their colour and size. Jumbled up together, dahlias can feel too busy, but if you divide them into three colour families, it's hard to go wrong.

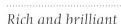

Rich and brilliant

This is my number one palette, my first love, the colours of stained glass. This includes red, orange and purple, with some almost-black ones mixed in to add sobriety.

Fashion is changing, but until recently the dark crimson and blacks were the truly chic dahlias that everyone wanted to grow. These dark-flowered forms prevent this otherwise jazzy lot from ending up looking like a bag of boiled sweets.

There's black, but no white in this group of flowers, not even in their centres. That's important in the colour mix: white is a powerful colour and jumps to the fore. With these intense, saturated colours as the backdrop, it would steal the show.

There's nothing better for a party than filling every surface you can spare with coloured glass vases, each one with a contrasting single-stem dahlia from this group. The very spiky ones (the Cactus or Semi-cactus types) have the strongest architectural presence, holding their own with needle petals pointing this way and that. But it's good to vary the flower form to get a full constellation of shape. Cut at different heights and with as much of a mix up of colours and shapes as you dare, it can look like your own table-top firework display (see page 374).

1 *Dahlia* 'Bishop's Children' (Single-flowered)
This is the only one here grown from seed, rather than tubers, which makes it cheap and easy to grow en masse. I love its single flowers, which are good with the compact *Verbena rigida*, both in the garden and in vases. They make a jewel-like, long-flowering duo, often still looking good in early November.

2 *D.* 'Bishop of Llandaff' (Paeony-flowered)
We grow our dahlias so that at least a third of them can provide nectar late into the year. If you can see the middle of the flower, some pollen and anthers, then it's a safe bet that the butterflies and bees can see it too. All the Bishop varieties fit the bill. We grow lots of them, including this classic scarlet one – I don't think you can better it for a true red.

3 *D.* 'Black Jack' (Semi-cactus)
Huge heads with spiky shapes, as well as tall stems that provide great scale. I love growing this in a huge pot in the Oast Garden, where the flowers tower over our heads.

4 *D.* 'Gerri Scott' (Semi-cactus)
A fantastic, newish and very spiky purple dahlia. This looks more like a rockpool sea anemone than any other dahlia I've grown, and I love it for that. It has a good vase life and looks spectacular cut for any arrangement, large or small.

5 *D.* 'Lou Farman' (Miscellaneous)
This is our most recently bred dahlia and it's named after my long-standing business partner and dear friend. It opens rich pink and fades elegantly.

6 *D.* 'Mel's Orange Marmalade' (Fimbriated)
Hard to beat for an orange. Its petals look as if they've been cut with crimping shears.

7 *D.* 'Purple Haze' (Miscellaneous)
Luscious, cardinal purple with rich crimson staining over the back of the petals.

8 *D.* 'Rip City' (Semi-cactus)
My first love, in deep, rich Bournville chocolate-crimson, with great texture and large flowerheads. It's quick-growing and vigorous and has a medium-length vase life.

9 *D.* 'Waltzing Mathilda' (Miscellaneous)
One of my top five overall dahlias, this is a rich and sultry semi-double, so good for pollinators. It has dark foliage and lasts well in a vase.

Soft and warm

My next colour palette is soft and warm. These have a large percentage of white in their tone, warmed up by orange, amber, rust, salmon pink or gold. This includes the delicate 'nude' colours, as well as milky coffee, blush, apricot, faded coral and peach. I think of these as rich tones, hugely diluted. They remind me of cashmere jerseys: comforting, pretty and romantic.

Lots of us are moving towards these softer colours in our dahlias. They look even better with a dab of dark crimson for contrast, or a splash of coppery bronze – both prevent the mix from becoming too sickly sweet.

1 *D.* 'Burlesca' (Ball)
So the huge flowerheads are in, but so – counterintuitively – are the miniscule: the petite Ball and Pompon types. Most people used to think of these as too neat and uptight, but again, that's all changing. 'Burlesca' is, in my view, at the top of the Ball group. It is a peachy flushed crimson and looks like honeycomb.

2 *D.* 'Café au Lait' (Decorative)
People are obsessed with 'Café au Lait' and it is miraculous for its range of colour from milky coffee and ivory to soft pink. The flowers are huge, so this is popular with bridal florists, as you've got a bouquet with only three heads.

3 *D.* 'Kelsey Annie Joy' (Collarette)
I played with Fuzzy Felt as a child, and as you can imagine, I usually used it to make flowers. They looked just like this.

4 *D.* 'Labyrinth' (Decorative)
My long-standing favourite dahlia in the soft, warm tonal range with flowers in a mix of colours. The bulk of each petal is peach, intensifying at the heart into a rich pink and fading at the edges to ivory. And it has whopper flowers which are super-pretty and curvy, with a lovely overall shape.

5

5 *D.* 'Molly Raven' (Decorative)
In 2017, we found 'Molly Raven', possibly my favourite of all, with petals that remind me of Venetian marbled paper. It has handsome ebony stems and green-washed crimson foliage.

6 *D.* 'Penhill Watermelon' (Decorative)
Almost peony-like in its luxurious curviness and curls. It's also a bountiful flower cropper, unusually so amongst large-flowered varieties.

7 *D.* 'Perch Hill' (Decorative)
We found 'Perch Hill' after trialling it from a breeder. My husband Adam picked one from our garden trial and brought it in and put it in a jug on the kitchen table. It lasted a week and I loved the succession of colours it went through, starting sort of ruby pink, moving to a soft brown-pink that was like calamine lotion mixed with mushroom soup. That sounds disgusting, but it's a uniquely beautiful dahlia and lasts longer in a vase than any I've grown.

8 *D.* 'Sarah Raven' (Anemone-flowered)
I fell on this one in a Dutch trial field in 2016. It has a crimson heart which fits well in the 'rich and brilliant' clan, but has a terracotta petal edge that makes it slot in perfectly with the soft and warm palette. It's incredibly easy to use in almost any mix. It is one of the first to flower (reliably by the middle of June here), has a better than average vase life and superb, straight, ebony stems, which make it perfect for cutting.

9 *D.* 'Schipper's Bronze' (Single-flowered)
I first saw 'Schipper's Bronze' with my Dutch dahlia-loving friend Dicky Schipper in a trial field back in 2015. I fell on this utterly marvellous dahlia, the colour of rusty tin with a bit of conker-brown mixed in, and it's become my favourite for a container.

10 *D.* 'Totally Tangerine' (Anemone-flowered)
Another excellent container dahlia in a soft, terracotta-orange, with a long flowering season. It's the perfect size for the front of a border or a large pot, which is where we always use it, but it does not cut well.

Sharp and cool

The third and final palette is this smart, crisp-coloured crew. Again, there is lots of white here. Some are pure white, some just have the lightest dash of another colour – perhaps primrose or pink or a touch of mauve. Blue-purple is the other key colour and there are also some star-burst shapes.

1 *Dahlia* 'Blue Bayou' (Anemone-flowered)
The Anemone-flowered dahlias are brilliant for pollinators. When I first grew 'Blue Bayou', it looked to me like the central tubes would be empty of nectar, but in fact it's the opposite. If you sit and watch bees and butterflies visit your dahlias, you'll see the Anemone-flowered seem to be the most attractive. Their flat plateau-like flowers are perfect landing pads, and even on a breezy day, butterflies can hold fast and have a good meal. Right into October, we have red admirals all over the 'Blue Bayou', and with butterflies increasingly rare, it's a really cheery sight.

2 *D.* 'Emory Paul' (Giant Decorative)
It's the football-sized 'Emory Paul' that beats them all in terms of scale. This stops you in your tracks. I can't believe the size of these every July or August, when they first open. Not a big flower producer, but crazy and marvellous.

3 *D.* 'Evanah' (Decorative)
A new-found favourite. Its flower looks as if it's been moulded from icing in soft pink and white with an angular, chiselled shape.

4 *D.* 'Fleurel' (Decorative)
I love the vast white lotus flower of this dahlia, which looks best floating in a bowl or shallow vase, simply on its own.

5 *D.* 'Hollyhill Spiderwoman' (Miscellaneous)
A bonkers explosion of a flower, which not only looks good, but produces more flowers than any other dahlia we've ever grown and is excellent for picking. But it does tend to wear itself out. We find it's one of our only varieties which needs replacing most years.

6 *D.* 'Honka Fragile' (Single Orchid)
'Honka Fragile' has done well in our trials, it's not fragile at all and it makes a surprisingly long-lasting cut flower. It's tall, but not too tall, so works well in a large container if staked. It's a Catherine wheel of a dahlia.

7 *D.* 'Small World' (Pompon) This white mini globe dahlia is a true drumstick, excellent when grown or arranged with the really big dahlias such as 'Café au Lait'. I love the size contrast. Many of the varieties in the Pompon and Ball groups (and the whites most of all) are popular with florists because they make good wedding buttonholes and small Victorian-style tuzzy-muzzies. They also tend to have the longest vase lives.

Natural slug and snail prevention

There's one issue that puts some people off dahlias, and that's slugs and snails. Dahlias are hugely attractive to slugs and snails, which love to feast on them, particularly in the spring. This is most likely to be an issue when dahlias are planted in mixed shrub and herbaceous beds, where there's plenty of cover.

We have, until recently, treated our densely planted permanent borders at Perch Hill with a biological control of nematodes that is effective against slugs, and to a lesser extent, snails. Watered on liberally when the dahlias emerge in mid-April and then again in mid-May, it seemed to do the trick.

We are, however, evolving a new system throughout the garden which I could not be more excited about. A tall, quite elegant grass called violet millet (*Panicum miliaceum* 'Violaceum') had self-sown through one section of the rose garden a few years ago. It might have come in via a wild bird food mix – it was right outside my office, where we have quite a grove of bird feeders, so it seems likely. As soon as the seed was ripe, the place was a flutter with goldfinches, and whenever I opened the door to go into the garden, it felt, for that week or two, as if I was in a natural aviary.

It was a lovely sight, so for that reason more than any other, we packed that bit of the garden with this millet together with its smaller relative, the delicate and wonderful cut flower, *Panicum capillare* 'Sparkling Fountain'.

Now, in August and September, the place is full of different members of the tit family. Joining the finches are blue tits, great tits, coal tits and even long tail tits, all there in good numbers every day. Then the thrushes come, and the robins and blackbirds, mainly feasting on the seed dropped below.

An unexpected and welcome side effect of this explosion in our wild bird population was an almost total wipe out of slugs and snails. So we tested this in the dahlia trial garden, interplanting the dahlias with hawthorns (for their berries), and lots of grasses, seeds and grains that attract birds.

Dahlia 'Cornel Brons' under one of the common hawthorns lining the drive.

We put in amaranth and heavy-seeded sunflowers, and the pink ornamental *Chenopodium giganteum*, as well as *Atriplex hortensis* var. *rubra* – all things we'd seen wild birds feasting on elsewhere. Teasel is a great plant for finches – chaffinches and goldfinches both seem to love it – but we can't take the risk of it self-seeding everywhere. Instead, we encourage it to grow in the wilder bits of the garden, where it can do what it likes. But together with cotton thistle (*Onopordum acanthium*) – an impressive, elegant, fifteen-foot, spiky monster that's another bird favourite – it needs to be kept under control.

The more containable, pretty grasses and grains can self-seed, and we've now rolled out patches of these different bird-friendly plants all over the garden. With that, and our huge tit and finch population explosion as a result, the pesky molluscs have become a rarer and rarer sight at Perch Hill. I don't know if these plants could quite replace our bird feeders (we have lots of these), but the two together seem to be working well – and are so much nicer for both us and the birds than slug pellets scattered everywhere.

Below *Panicum miliaceum* 'Violaceum' makes a feast for small garden birds once ripe (particularly goldfinches), with dahlia 'Belle of Barmera' and 'Sarah Raven' in the dahlia trial.
Opposite Homemade willow and dogwood frames holding fat and seed cylinders that hang in many of our trees throughout autumn and winter.

Practical
September

September, just like August, tends to be filled with small jobs, rather than one massive campaign.

Cosmos are still going strong – I try to spend a peaceful half hour deadheading them, and I do the same with dahlias. I am repaid with hundreds more flowers over the next few weeks.

We start sowing hardy annual cut flowers and garden fillers early in the month. It gives us bigger, better plants that can flower a good six weeks earlier than those spring sown. In fact, some varieties are utterly transformed by autumn sowing. *Ammi majus*, cornflowers ('Black Ball' and 'Blue Boy'), *Euphorbia oblongata* and scabious ('Tall Double Mix' and 'Black Cat') all form fantastically healthy, floriferous plants, almost twice the size of their equivalents sown in the spring. Honeywort can be in flower by March if we sow it now, and we often add *Linaria maroccana*. (See undercover sowing instructions on page 80.) We also direct sow opium poppy 'Black Peony' outside.

Any hardy annuals sown early in September will need planting out within about four weeks – the sooner the better, once they reach about 2cm (1in). They need to get their roots down while the soil is still warm, or they won't survive the winter. We really see the benefit of all this when they flower four to six weeks earlier the following spring.

It's also vital to get all our biennials in place now. They were sown in June (see page 191), and they also need to get their roots down well before flowering.

At this time of year, I often have a hunt round the garden for self-sown seedlings of hardy annuals and biennials such as English marigolds, honeywort and honesty. I make sure they are not overcrowded where they've sprung up, and if they are, I transplant them, spacing them at 30cm (12in) apart in a good sunny patch (or shady area, in the case of honesty). I often create a line of them in a cutting patch to harvest flowers the following year.

Planting Early Bulbs

Anemone coronaria

Garden anemones benefit from being planted earlier than most bulbs. There is a newly bred lot of these anemones that comes from Israel (where they grow wild), in a good range of colours. My favourites are 'Jerusalem Blue', with big, rich purple-blue flowers that have a velvety texture, as well as 'Galilee Red'. We grow some undercover and some outside, where they are perennial.

- Soak corms in a bowl of water for a couple of hours so they plump up.
- Plant some straight in the ground and have cloches ready to protect them if we have severe weather.
- In case we have a very cold winter, we also plant a few individual corms into 1-litre pots and store them in a greenhouse or cold frame until February, when they can be planted out with well-established roots.

- We also plant a handful into our greenhouse beds to ensure really early flowers.
- Plant them shallow, placing them claws upwards, about 5cm (2in) deep and about 10cm (4in) apart.
- Water them well on planting and then leave them be. If growing in the greenhouse or in pots, check occasionally that the soil is not bone dry, but don't overwater. They'll rot if too wet, but need moist soil.

Sowing violas

We often grow our anemones with pansies and violas (such as *Viola* 'Sorbet Phantom', see page 38) and it's best to sow those seeds now. Violas have tiny seeds and we don't want huge numbers, so sow them into a half-sized seed tray (see page 80).

Forcing Hyacinths

If I want hyacinths for Christmas, I get going with them now. Unprepared bulbs (ones that haven't been pre-chilled) take about fifteen weeks to flower, so I try to plant a handful in a pot (or a bowl with added drainage) from the beginning of September and every fortnight until mid-October. I will have put some bulbs in the fridge in August (see page 263), and these ones are now 'prepared' and only take ten weeks from planting to flower.

The cold fools them into thinking it's winter, so stimulating a biochemical response inside the bulb to make it start flowering. The dark gives the root time to develop before the light pulls the flower and leaves from the bulb. Vita Sackville-West put them under her bed, but famously lived without heating!

One note of caution: hyacinth bulbs can irritate the skin, causing an itchy sensation for some people. So wash your hands well after planting or wear gloves.

- If you like the idea of bulbs in your ceramic salad bowls (but like eating salad from them and don't want to lose all your favourite bowls for months), there's an easy solution. Find a plastic bowl (a cheap mixing bowl works well) that fits just inside the ceramic bowl and plant the bulbs in that. When the hyacinths are almost in flower, slot the plastic bowl into its decorative home.
- All bulbs need to be grown in a well-drained but moist environment with a soil structure that is strong enough to anchor them. That's often the problem with cheap shop-bought forced bulbs: they are planted in peat-based compost, which when it dries out becomes as light as dust and can't hold the elongating stem and top-heavy flowerhead. At Perch Hill, we can use molehills – they give us the perfect, crumbly, fine consistency soil.
- With no access to molehills, use John Innes No. 3 or a specially formulated bulb compost. To this, add some grit. To a bucket of molehill soil, add just under half a bucket of grit. If you're using John Innes No. 3, use a ratio of 1 part grit to 3 parts compost.
- Add 2½cm (1in) of pure grit in the bottom of

Sowing hardy annuals including winter-flowering pansies; here *Viola* 'Antique Shades Mix'.

every pot, then add a shallow layer of the compost and grit mix. Then add the bulbs with the pointy tip just below the soil surface. Fill in around those with the compost mix.

✿ If you are using a contained pot with no holes in the bottom, use bulb fibre instead. This is expensive, but contains charcoal and grit in the right proportions. The charcoal is porous, and with the grit, it will keep the soil damp but not too heavy. This prevents rot and fungal disease.

✿ Hyacinths need fifteen weeks at a

temperature of around 10°C (50°F). If you're using prepared bulbs, they need ten weeks at this temperature. For that period, they also need to be in the dark, so leave them in a cellar, cool cupboard, potting shed or garage.

✿ Once the sprouts are up to a good 2-3cm (1in), bring the pots in somewhere warmer (above 15°C/60°F) and they will start to sprout quickly. They think spring has arrived and will be in flower in 2-3 weeks.

✿ To stop the stems and leaves flopping about, create a nest of twigs to support them (see the plant supports on page 88).

Forcing Amaryllis

If you plant *Hippeastrum* (commonly known as amaryllis) towards the end of September, it will be in flower in eight to ten weeks. One single bulb should then go on looking good for about the same length of time. See page 359 for ideas on how to use your amaryllis once they're growing and starting to flower, page 356 for my favourite amaryllis varieties, and page 368 for how to store the bulbs.

* These South American natives do not require a cold spell as the hyacinths do.
* Before planting, hydrate the desiccated roots by soaking them in tepid tap water overnight. The easiest way to do this is to rest the base of the bulb on the rim of an appropriately sized jam jar so that all the roots (but not the bulb base), can sit in the water below.
* Amaryllis like a tight-fitting pot, with about 2½cm (1in) between the bulb and the side of the pot. The pot size will depend on the age and size of the bulb you're planting, but as a rough idea, for one bulb you'll need a 15-20cm (6-8in) diameter container with a depth of at least 30cm (12in). This will allow plenty of room for good root growth.
* One bulb looks good, but three or five look even better and can hold their own for a good three months. Their lush, flat and iris-like leaves remain a good bright green before there's any sign of a flower bud. If you plant several, bear the same tight planting rules in mind, leaving 3-4cm (1½in) between each bulb.
* Amaryllis have a tendency to rot, so drainage is vital. Put a good handful of crocks in the bottom of your pot to help with drainage.
* You can use potting compost, but they like their soil rich, so a mix of 1 part well-rotted manure, 1 part grit or sand, and 2 parts leaf mould is even better.
* In the wild, they root quite superficially, with about one third of their huge bulbs poking out of the soil. This allows heavy rain to flow away from the crown of the bulb easily and lessens the likelihood of rot. You should plant them like this in a pot. It's important that the shoulder of the bulb sits a third above the surface of the compost. It's the apex of the bulb (where the leaves emerge) that is the most vulnerable to rot and where water can seep in and decay the heart, so this part mustn't sit wet on watering. It might look odd, with the bulb perched high, but that's how it should be.
* Planted so high, the whole thing is in danger of being top-heavy and toppling over as it grows. They grow to at least 45cm (18in), depending on size of bulb and variety, so you will need to give it some support. You could just tap a cane in and tie the stem to that as soon as it gets to a decent height. But if you've planted more than one bulb, it's good to go one step further. Poke in twigs (ideally ones that look good, such as silver birch, dogwood, hazel or alder) and push some in around the edge and a few branches in between, taking care not to pierce a bulb as you go. The nest looks good and provides support.

❦ These are hot-country tender plants and they love the warmth. Place them in a light and well-ventilated spot, free from draughts, somewhere that's about 20°C (70°F). A shelf above a radiator is ideal.

❦ Keep the compost moist until a shoot appears and then water more frequently, about twice a week. Water from the top using tepid tap water, not from the bottom, and once the water has drained through into the saucer, tip it away. Don't overwater.

❦ After the plant begins to grow, feeding is essential. You can add a complete slow-release fertiliser to the potting medium when you plant, or use a liquid fertiliser twice a month when in flower. Either works fine and will ensure you have an even bigger, better bulb to dry and store for planting again next year.

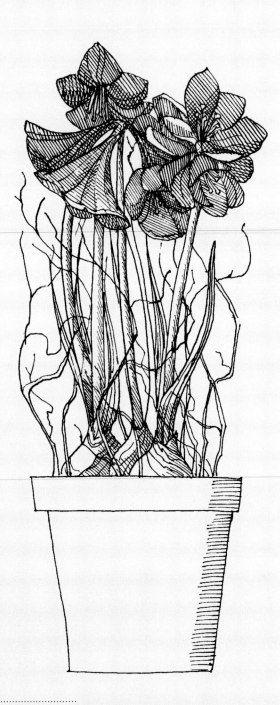

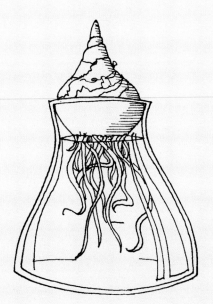

October

Perch Hill is surrounded by woods, giving it a feeling of enclosure. The farmhouse sits in the middle of its cleared fields and in autumn the atmosphere is dominated by trees. There's field maple, wild cherry, hornbeam and ash, but above all the oaks, the leaves now turning to turmeric, ochre and bronze so typical of the Sussex Weald.

October is the opposite to March. Everything is slowly pulling back in, the garden back into the soil and the grass into the fields. In the woods, there's deep coppery leaf litter without a spot of green. Pigment drains away, even in our jungle of dahlias the petal colours are softening.

We don't usually get frosts in October, but we do get plenty of wind and rain, so even with all the staking we do, the cutting gardens start to feel a bit lopsided with stems breaking and hanging. There's still plenty to pick though. Cleomes look fresh and the cosmos are usually still going strong, with zinnias beside them, fading now, the stalwart small-flowered forms such as *Zinnia haageana* 'Jazzy' flowering the longest.

Good old verbenas are happily pumping away, none better than the compact, bright purple *Verbena rigida*, which we use as path and border edging all over the place. That looks as bright now as it did in July. And I adore nasturtiums; they have a big presence here in October. Even with heavy picking, for both our salads and flower arrangements, I am surprised every year by how long they keep going.

It's right now that chrysanthemums reach their peak – and I couldn't feel more grateful. Just as the dahlias start to turn a little brown from the rain, chrysanthemums come into flower. If, like me, you love flowers in the house, they are a godsend.

Trained on the walls and in pots in sheltered places, the same can be said of abutilon. I love the party-dress look of these flowers, especially with the new muted ones, such as 'Love and Kisses'. We will soon bring them into the greenhouse where they continue to flower right through the winter until we cut them back, but they have great presence for now in the Dutch Yard.

We have the odd exceptional rose still in flower, such as 'The Simple Life', 'Lady Emma Hamilton' and 'Belle Epoque', but certainly towards the end of the month it's the salvias we have to salute when it comes to long garden service. They are all miraculously full of verve and vigour, covered in flowers from the ground up. Whether it's the compact *S. microphylla* and *S. greggii* types, which we have mainly in the rose garden (see page 184 in June), or the taller, bushy forms, such as S. 'Amistad', *S. guaranitica* 'Black and Bloom' and *S. involucrata* 'Hadspen', they're still looking wonderful. Penstemons such as 'Pensham Just Jayne' and the traditional 'Garnet' are still flowering, but not even they can compete with the range of colour, texture and scale of the salvias.

Below Our edible garden with *Verbena rigida*, Thai and cinnamon basil, and dahlia 'Bishop's Children' lining the paths.
Opposite An October urn of soft and warm chrysanthemums, including 'Pandion Bronze', 'Salmon Allouise' and 'Avignon Pink'.

Salvias and scented-leaf
pelargoniums still fill the beds in
the Rose and Herb Garden in early
October, with pot-grown chillies and
our second sowing of Greek bush
basil about to come in so that
they're out of the frost.

We have bog sage (*Salvia uliginosa*) too, one of my garden favourites because of its uniquely coloured flowers, a true Aegean-sea blue, exactly what I feel I need right now. For the boldest contrast and a really cheery zap of colour, we have them planted with *Nerine bowdenii* in a narrow border on the south-facing side of the house. Between them, these two give colours so tropically intense you'd hardly believe you're in a garden in Britain in June, let alone October.

There's been lots of breeding of nerines to make the ones with saturated colour hardier, and we've been trialling a few, as well as some amarines, which are a cross between *Amaryllis belladonna* and *Nerine bowdenii*. Unsurprisingly, given the parents (which are both slow to settle in), they've yet to flower much, even in their third year, so I don't know if these are going to be a success. But with the possibility of that level of colour in October, they're certainly an exciting prospect.

So there *are* plants that give floweriness and colour as the growing season draws to a close towards the end of the month, and we try to find room for as many here as we can, to guarantee a garden that sings right to the end.

Opposite Nerines and salvias (*N. bowdenii* and *S. uliginosa*) give guaranteed colour brilliance in a sheltered spot late in the year. Below *Rhodochiton atrosanguineus* on a silver birch teepee with salvia 'Blue Note' giving substance below.

Salvias

I've already written about our passion for the small-flowered, small-leaved salvias and their use as companion plants to roses (see pages 171 and 184), but I can't think of the garden here in October and fail to return to salvias once more. We have so many still looking fantastic, filling our pots and borders with colour.

We first got into salvias here as a way to feed pollinators late in the year, as from July onwards there's a UK-wide lack of forage. Plants such as single dahlias and late-flowering salvias are just the thing. Both pollen- and nectar-rich, they are important plants for all of us to grow. Stand next to a clump of salvia in October and you'll practically be deafened by that optimistic, noisy buzzing.

Native to warmer climates, half-hardy herbaceous perennial salvias are programmed to have exceptionally long flowering seasons. Even when grown in our climate, this is still the case. We've had them out here in snow and they survived with no extra fleece or paraphernalia to protect them. And so many of them seem impervious to strong winds or torrential rain.

Perhaps the king of the lot is *Salvia* 'Amistad', which enjoyed a great rise to fame as soon as it was introduced in 2012, and rightly so. It's covered in blue-purple velvet glove-like flowers from June until November here, and gets bigger and better year on year. The brilliance of 'Amistad' is that it looks good for so long, and we're particularly reliant on it right now, when almost everything else is starting to tire. It's lovely in July with phlox and crocosmias, but there are other things then to carry the colour baton; it's now, late in the year, that its vibrancy is so welcome.

Salvias still flowering away at this time of year. From left to right: 'Black and Bloom', *S. confertiflora*, 'Cerro Potosí', 'Phyllis' Fancy', *S. leucantha*, 'Stormy Pink', 'Hadspen', *S. curviflora*, 'Jezebel', 'Love and Wishes' and 'Amistad'.

As well as the rich blue-purples that the salvia genus is known for, I really love the deep pinks such as *S. involucrata* 'Hadspen' and *S. curviflora*, each flower as velvety as can be. We have both these planted with *Rosa* × *odorata* 'Mutabilis',

still covered in flowers and buds in October, and all contrasting to a zap of 'Amistad'. All three of these salvias are perennial with us, re-emerging every spring from plants put out several years ago. They are, strictly speaking, half-hardy, but with lots of grit added to their planting position and mulched deeply with 15cm (6in) of municipal green waste over the crown, they're fine here. Their aerial parts, the stems and leaves above ground, will succumb to frost, but their roots will survive even harsh winters as long as they are on well-drained soil. We tend to do cuttings as insurance (see page 267), but we haven't needed them to replace any frost-killed plants, so we just end up with more and more salvias.

Below The Perennial Cutting Garden is still pretty full of flowers in October, with salvia 'Amistad', *Stipa gigantea* and *Cobaea scandens* growing over the arches, giving us flowers to pick.

Opposite *Salvia involucrata* 'Hadspen' takes up the rich pink colour baton from rose 'Charles de Mills' in a central bed in the Farmhouse Garden, which looks good until Christmas in a mild year in this sheltered spot.

Best of the salvias

See June for *S. microphylla* and *S. greggii* types

For a colourful garden from June until at least October, look no further. If I had a small sunny courtyard garden and wanted plants which looked good on minimal TLC, I would choose from these.

1 *Salvia* 'Amistad'
In terms of overall virtuoso performance, *Salvia* 'Amistad', with its ink-black calyx and indigo-purple flower, steals the show. It flowers from late May and is just as covered in bud in October. It's a truly marvellous plant that no garden should be without.

2 *S.* 'Love and Wishes' and 'Ember's Wish'
Totally exceptional, smothered in flower from June to November. Velvety, vigorous and handsome. 'Love and Wishes' pictured here.

3 *S.* 'Phyllis' Fancy'
In a fashionable muted blue-grey, the felted stem and calyx of this salvia is very Farrow & Ball, in contrast to its slightly paler flowers. This one is the perfect scale for summer containers, on its own or mixed with the slightly richer-coloured African blue basil. They look good together remarkably late in the season.

4 *S.* 'Salmia Dark Purple' and 'Salmia Pink'
I run out of superlatives for these two ('Salmia Dark Purple' is pictured). Related to the 'Wish' series (see 'Love and Wishes'), these are perhaps even better and longer-performing, with slightly larger flowers and more of them. A pair of truly exceptional plants for June to November. The calyces give remarkable garden presence even after the petals have fallen, and we only finally relinquish these two when they have to be chopped down to make way for tulip planting in November or even December.

5 *S. confertiflora*
With scarlet-orange flowers and bold, crinkly leaves, this is upstanding whatever the weather throws at it. We grow it in borders and pots, so it can be moved inside for winter architecture in the greenhouse; it then flowers there almost all year.

6 *S. curviflora*
Delicious velvet glove-like flowers in the richest pink. I love this one intermingling with *S. involucrata* 'Hadspen', with its matching (but chunkier) velvety flowers. It's classified as a half-hardy woody perennial, but once established, survives the winter fine here.

7 *S. discolor*
From Peru, and blacker than chocolate cosmos and dahlia 'Karma Choc', this looks like its flowers have been cut from silk velvet, a hugely glamorous contrast to the bright green, washed-silver calyx. It's another one for the greenhouse, but it can come out to sit somewhere sheltered and warm in summer.

8 *S. guaranitica* 'Black and Bloom'
A true blue, very similar to 'Amistad', but genuinely blue, rather than purple. It is also said to be hardier for more exposed sites, but as both this and 'Amistad' do fine here, I can't vouch for that.

9 *S. involucrata* 'Hadspen', 'Boutin' and 'Bethellii'
These are the biggest and boldest of the late-flowering crew with balloon-like calyces in deep pink ('Hadspen' is pictured).

They're super-showy and reminiscent of a lotus flower. These come in three almost interchangeable shades of nerine-pink, 'Bethellii' possibly a little richer in its tone.

10 *S. leucantha* 'Purple Velvet'
A really bright purple variety that has a strong colour with fantastic super-soft texture. It's a 'short day' salvia that starts to initiate flower buds as the day-length decreases. I find it's best grown in a large terracotta pot and can stay outside until late October, then I bring it into a frost-free glasshouse or conservatory. It will flower from September through to May, when increased day-length allows it to revert to vegetative growth mode.

11 *S. patens* 'Cambridge Blue', 'Lavender Lady' and 'Guanajuato'
Mainly grown from seed, rather than cuttings, the *S. patens* types are classics for late summer and autumn. 'Guanajuato' is a rich royal blue, 'Cambridge Blue' (pictured here) as it sounds, a soft sky hue, and 'Lavender Lady' a very similar soft and pale shade, but with a touch of mauve.

12 *S. uliginosa*
Flowering August to October, this romps when it's happily situated, but a perfect site is not always easy to find. It likes a moist soil, but also heat. We have it planted in the shelter of a south-facing wall here (with nerines) but it needs a good mulch.

Nasturtiums

Nasturtiums (*Tropaeolum*) are powerful-looking plants with great trumpet-like flower shapes that are strong and noisy, but with the right varieties, not coarse. They're invaluable for performing late in the year. They're classified in seed and plant catalogues as half-hardy annuals, but sown in summer, they're able to put up with some cold nights here and keep flowering through to at least the end of October. And they tend to recur from one year to the next, self-seeding happily.

I'm sure many of us can bring to mind that famous view of Monet's garden, looking up at the house, with waves of nasturtiums undulating across the path from both sides. I love that image and have copied the idea here many times. The trailing nasturtiums are the ones we mainly use. They sprawl so vigorously that they can be a bit of menace in a crowded flowerbed, where their triffid arms tend to overwhelm delicate plants. But over gravel, brick or stone, they thrive, loving the heat while softening the hard lines.

The very bright and cheery semi-double 'Jewel Cherry Rose' is a good path edger. I love its brilliant coral flowers alternating with the royal blue *Anagallis monellii*, and with Spanish flag climbing up above. From a spring sowing of nasturtiums, we can get this going for summer, but by sowing them in June or even July, we ensure this colour ribbon edges the paths into late autumn.

We also sow nasturtiums to fill empty patches quickly. There's a tricky area at the end of our dahlia trial garden that has a sunken water tank below it, with only a few inches of soil before you hit the concrete. It's been a bit of a blank spot over the years. We've tried filling it with dahlias in pots, but it's too far from a tap for easy watering. Then we tried cosmos, but they struggled with the lack of soil depth. Nasturtiums are one of those plants where the poorer and thinner the soil, the better, so they are an ideal candidate for exactly this sort of place. Last summer and autumn we mixed in a new range of pineapple flowers, namely *Eucomis* Aloha series. They looked good together and really lasted.

Moroccan bowls full of nasturtium 'Purple Emperor', 'Jewel Cherry Rose', 'Black Velvet', 'Empress of India' and 'Gleam Salmon'.

Another nasturtium we grow lots of here is the dark-flowered and dark-leaved 'Empress of India'. It's not only a sprawler, but it climbs a bit too. I love the way it spreads out in every direction, like the hands of a clock. When one arm hits a wall or fence, it just moves vertically to create elegant shapes. We have a rabbit problem, particularly down the drive, so our chestnut paling there is lined with chicken wire, but it becomes invisible with these nasturtiums clambering over it.

Every part of the nasturtium is edible: the flowers, young leaves and seedpods are all delicious, with a strong, peppery, rocket-like taste. The pods in particular are good for pickling and we have used them as the home-grown equivalent of capers. You can't grow capers in the UK – I've tried – as we don't have the necessary length of growing season, but nasturtiums make a good substitute.

As an edible flower, we grow lots of the new nasturtium 'Bloody Mary', which stands out so well on any plate. For the same reason, we love the deep red-brown 'Tip Top Mahogany'. With this, we pick its contrasting-coloured bright green leaves as well as the flowers. It looks wonderful in the garden and mixed together in food. Pretty much all nasturtiums are perfect for late-summer salads, but you can also use the flowers and leaves in fish cakes. Just add them instead of black pepper (the flavour is similar) to the fish and potato mix and they'll wilt.

Nasturtiums are famously useful companion plants, attractants of good bugs such as lacewings and ladybirds, as well as a buffer for bad. With their strong smell, they draw cabbage white caterpillars away from your cabbages, and as part of the brassica family, they help divert flea beetle from salad plants such as rocket and mizuna.

There has been a good deal of breeding recently, with new dusky colours being added to the bold and rich. In the last few years, we've been picking lots for late-season bowls and vases, searing their stem ends in boiling water first. They've become some of my favourite October picking flowers, with both the muted 'Ladybird Rose' and 'Purple Emperor' as front runners.

Best of the nasturtiums

Nasturtiums are classic cottage garden annuals and currently very on-trend, particularly in the new soft colour ranges. They are incredibly easy to grow in any spot in the sun and self-seed easily from one year to the next. They're a great plant for those just beginning to garden.

1 *T. majus* 'Bloody Mary'
Flowers vary from one plant to the next, primrose yellow with a crimson splotch and delicate veining, to almost completely crimson with contrasting yellow markings. We use this more than any other for salads as it really stands out in the bowl.

2 *T. majus* 'Empress of India'
This has dark stems with dark green-washed crimson leaves, with new growing tips a sort of slate-crimson. That's all in brilliant contrast to rich orange flowers.

3 *T. majus* 'Gleam Salmon'
Newly bred, apricot flushed at the petal heart with orange-crimson, this is a marvellous cut flower for mixed bunches and on its own, with the best-ever curves and shapes to every flower and stem.

4 *T. majus* 'Jewel Cherry Rose'
The brightest of our nasturtiums, with semi-double flowers in scarlet-crimson. It's a relatively vigorous trailer, not a climber.

5 *T. majus* 'Purple Emperor'
An exciting new variety that starts a smoky crimson and fades in the most beautiful way into a sort of sepia brown. Another trailing variety.

6 *T. majus* 'Tip Top Mahogany'
This is one of the nasturtiums we've grown here for longest as I love the strong contrast between leaf and flower colour: the leaves are the brightest acid-green and the flowers a velvet crimson-orange.

7 *T. minus* 'Ladybird Rose'
Subtle and pretty, the flowers are crimson over calamine mixed with café au lait. This is from a different branch of the family, *T. minus*, and more delicate with it. I love it as a cut flower and it's a good container plant for an ornamental veg garden or window box.

An October wreath with hydrangeas ('Madame Emile Mouillère', 'Little Lime' and 'Wim's Red'), together with *Panicum capillare* 'Sparkling Fountain' and my favourite crab apple, *Malus* 'Dartmouth', surrounded by the statuesque *Salvia confertiflora* and the turning leaves of amelanchier.

Chrysanthemums

I've long been a fan of chrysanthemums and their late flowers, which give us great towering vases that last over two weeks. With the right varieties, they can be beauties, but I've felt a bit of a lone voice in my love for them over the last twenty years.

That has now all changed. I knew the tables had turned for chrysanthemums when a few years ago I walked into the Brooklyn workshop of Saipua (now no longer there, but then a hyper-tasteful and super-fashionable New York florist) to be greeted with buckets and buckets of them. These weren't the old-fashioned, prim types, with circles of flowers around a central golden disc, every petal so neatly placed they could almost be plastic. This lot were a very different crew, all in the most cadaverous tones. They cost a fortune and were destined for a flower installation at The Met.

As an ex-doctor who has spent far too much time on the fifteenth floor of Charing Cross Hospital doing anatomy dissections, the dead-flesh hues that are so fashionable can be a bit of a challenge, but in flowers it's strangely alluring and I fell for these chrysanths. A more polite description of the colour would be 'oyster', but if these pale shades are still too much, there are some less deathly tones in the ever-expanding chrysanthemum range.

Since gardeners all got hooked on dahlias, plant breeders have anticipated chrysanths would be the next big thing and have been producing some ground-breaking varieties. It's often the case that the best ones are bred for the cut-flower industry to start with and are under restrictive Plant Breeders' Rights, but in time the licences lapse and they become available as garden plants.

The old-fashioned, multi-headed, so-called 'spray' chrysanths are not usually objects of joy, whereas plenty of the new ones really are, with individual flowers the size of a hand, stacked with petals waywardly curving in and out. If I had to choose one flower I want more of going forward, it would be chrysanthemums.

No one could tell me the name of the variety I saw in New York, but I have found some good approximations. I've avoided those in the Singles group, where the centre of the flower is visible.

Chrysanthemum 'Avignon Pink' with *Nerine bowdenii* 'Ostara'.

These are the only ones that have forage for pollinators, but sadly they're not beauties. It's the doubles I want, where the central flower parts (the anthers and nectaries) are bred into extra petals. (To make it up to the butterflies and bees, we do a late sowing of something like honeywort or *Echium vulgare*.)

On aesthetics alone, you're pretty safe selecting from either the Spider or Decorative groups. The New York cadaver form was one of the latter, and as long as you find a colour you like, these all have bold, unfussy flowers.

My current favourite is the smoky mother-of-pearl 'Avignon Pink', marvellous in a vase cut on the longest possible stem and arranged on their own. They're also good picked shorter to mix with the similarly coloured *Nerine bowdenii* 'Ostara'.

A good range of bronzes is also available. 'Pandion Bronze' is a beautiful copper and it flowers for ages. I love the crazy-headed 'Spider Bronze', which is like a firework, ideal for cutting short and spreading out as a constellation down the centre of a harvest festival dinner table.

The dark crimsons are non-controversial. You'd be hard pushed not to like this colour, and where there's some conker-brown mixed

Below Chrysanthemums 'Pandion Bronze', 'Avignon Pink', 'Salmon Allouise' and 'Spider Bronze'. These all last over two weeks in a vase.
Opposite My favourite chrysanthemum 'Spider Bronze', looking brilliant in a jade glass vase.

An autumn urn of chrysanthemums ('Bigoudi Red', 'Tula Improved' and 'Allouise Orange') with scented pelargoniums ('Chocolate Peppermint', *P. tomentosum* and *P. quercifolium*).

in, so much the better. I love 'Bigoudi Red' for this reason, rich and delicious, like the best dark chocolate. But 'Tarantula Red' is my favourite of this lot, in the deepest velvet crimson, with flowers shaped like a starburst.

A colour that needs careful selection is salmon. Until quite recently I would have skipped straight over that section in any catalogue, but have realised that some of the softer versions of salmon, with variations in the petal tone, can be magical. Chrysanthemum 'Salmon Allouise' is a good example, a newish colour from the long-standing and reliable Allouise series, and I love it, with the gentle colour on the petal's reverse mitigating any fishiness.

Just like dahlias, chysanths are cut-and-come-again (though admittedly to a lesser degree). If you take the leader out, picking the top clump of flowers, the buds below will continue to develop, so you can pick a bunch again two weeks later. And you can do this from October when almost everything else is mushy and brown.

To get them looking their best, we plant three cuttings to a large pot (around 7-litres) and sink the whole pot in the ground (see page 267 for more on cuttings). As with almost all plants we grow for picking, we want them to be stocky, so after a couple of weeks we pinch them out, removing the top growth down to three or four leaves (so they're about 22cm/9in tall).

When the weather gets bad in October, we lift the pots out of the ground and bring them into the greenhouse to replace our tomatoes (a system of plant rotation which works well), and they will go on until Christmas with protection. With our increasingly long and mild autumns, you can get away without doing this, particularly if you have a sunny porch you can place them in, out of the autumn rain. But you'll get a larger harvest if you bring them undercover.

Best of the chrysanthemums

It's the more wayward, firework chrysanthemums I want to grow. They are utterly invaluable for flowering so late in the year.

1 *Chrysanthemum* 'Avignon Pink'
A soft, sophisticated and smoky, rather than sugary, pink with a lovely shape. I particularly love this variety.

2 *C.* 'Bigoudi Red'
Exceptionally long-flowering, often from August until November, even outside. It seems to put up with the weather well, so if you don't have anywhere undercover, this is the one for you.

3 *C.* 'Gompie Red' and 'Gompie Super'
These are spray types, which I tend to avoid, but I actually really like these two: one in crimson and one in a more purple tone ('Gompie Super' is pictured). They are neat, but not too neat, and very flowery over a long time.

4 *C.* 'Pandion Bronze'
Quite a standard, neat shape and flower form, but in a subtle and lovely colour. It's a true bronze, not orange (which lots of so-called bronze chrysanths turn out to be).

5 *C.* 'Shamrock'
Sharp lime green with a bold, spiky shape, this has been a favourite for ages as it's good in contrast to the dark purples, crimsons and mahoganies. It's a true sharpener.

6 *C.* 'Smokey Purple'
One of the earliest to flower with a good shaggy shape and three-week vase life. Ideal for the person who can only grow their chrysanths outside.

7 *C.* 'Spider Bronze'
Another favourite that looks wonderful cut and arranged on its own or with almost any mix. If I was to grow one for cutting, it would definitely be this – great colour and the best starry flower form.

8 *C.* 'Tarantula Red'
The richest crimson Spider type, with petals this way and that, and suitably crazy and un-plastic-looking. I love this growing in the garden and cut for a vase.

9 *C.* 'Tula Improved'
Josie, our head gardener, loves this one. The shading on the petals, the green centre and the whole starburst look makes it delicate, but it still has presence. There's a whole range of these starburst forms. We're trialling five different Tula colour forms this year.

Practical
October

October is the start of mass bulb planting here. It's our main mission at this time of year, but there are a few other things we must complete too.

Any hardy annuals sown undercover (see page 127 in April) and not yet planted out need to be put into the ground as the number one priority. It's almost too late for biennial planting (see page 191 in June for sowing), but if we have stragglers, planting them is now urgent.

There's a lot of tidying and putting things away. We bring pots of tender pelargoniums and echeverias undercover and also a few beautiful terracotta pots with ornate decoration that I've brought back from Italy and Crete – I'm not totally sure they're completely frost-proof, so we tend to bring them into our open-sided hovel or the greenhouse for protection. It's the twiddly decorations on Tuscan pots or the handles of Greek olive jars that seem to be vulnerable.

Getting outside and harvesting while we still have the chance is also a priority. Once we've had an autumn gale, it will be too late. We have a good scout for colourful seedpods and hips: Chinese lanterns (*Physalis alkekengi*) are looking their best and agapanthus (particularly 'Queen Mum', which seems to hold on to its pods longer than most) are like green faceted beads on strong wiry stems. We always try to pick lots of hydrangea heads, including from the climbing hydrangea (*H. petiolaris*), which we have growing on the north-facing side of the house.

It's always worth a big deadhead of dahlias: I cut all the spent flowers off to the buds below them. Given this bit of TLC, they'll keep on flowering a bit until the first hard frost. I pick a few pristine heads as I go. Cut short and arranged in small coloured glasses or bottles and placed all the way down the centre of a table, they give us pretty spectacular and instant party flowers.

While I'm at it, I deadhead all our tender perennials – pelargoniums, arctotis, argyranthemums and trailing verbenas – as this gives them a new burst of life. We combine this with a feed of liquid seaweed or general fertiliser as an extra tonic to keep them going that little bit longer.

Planting Bulbs in Grass

Our wildflower meadow is bang outside our kitchen window (see page 138) and we try to embellish it a little more each year, adding two or three new bulb species in large numbers to extend its performance.

October is the ideal time to plant lots of spring-flowering bulbs, such as *Crocus tommasinianus*, *C. vernus* and *C. chrysanthus* hybrids. It's also the time for scillas and miniature narcissi (see page 146 for my selection). There's no more delicately beautiful bulb than our native snake's head fritillary and now is a good time for planting. In our damp spots, we also plant camassia bulbs.

- ❀ Mow the grass.
- ❀ If you're planting two or three different bulb varieties, mix the lot up together in a trug.

❦ If you have a large area of grass, divide the lawn into squares (about 3-4 metre/12-13ft square) using canes around the edge as a reminder.

❦ Take handfuls of the mixed bulbs and throw them. That gives a pretty random distribution. Try to stand back at this stage and look at the whole lot before you start planting, and give their positions a tweak to make things look as natural as possible.

❦ Using a bulb planter, with a long handle like a spade, is the quickest and easiest way of planting bulbs in grass. It acts like a corer, removing a cylinder of soil, and it easily cuts through the fibrous flower and grass roots with minimal damage.

❦ The job of planting is much quicker with two people. One person digs the hole, the other adds a handful of grit (or spent compost) into the hole, then the bulb. Meanwhile, the first person is digging the next hole, dislodging the soil core, and the second person can pop that over the newly planted bulb, firming it down.

❦ Finish by mowing the grass again as late as possible at the end of October or even in November if the autumn is warm and wet. You could do this again in January if the winter continues to be mild. The bulbs will then show clearly through short grass the following spring.

Planting Alliums and Narcissi

It is allium and narcissi planting time. Both benefit from an extensive root system, so it's worth getting them into the ground this month, rather than after the first frost (as with tulips).

❦ Narcissi and alliums are tolerant bulbs. They'll grow well on a light, sandy soil in full sun and happily on heavy clay if you add lots of grit to their planting position. If the patch you're planting is sitting quite wet, use a small handful of pea shingle (we buy it by the ton from a builders' merchant) to line the planting holes.

❦ Narcissi will tolerate partial shade, whereas alliums prefer full sun.

❦ Unless they're whoppers, plant in clumps of at least 15. With fewer, you get a dotty effect.

❦ Use a bulb planter (long-handled is best so you don't hurt your back) to create a 15cm (6in) hole. This is the right depth for most of the bigger narcissi and alliums – about twice the depth of the bulb.

❦ N. 'Canaliculatus', 'Minnow' and 'Bell Song' are smaller, so only need about 5-6cm (3in) of soil above them.

❦ Line the newly dug hole with a bit of grit, then plop the bulb in pointy end up and cover with soil.

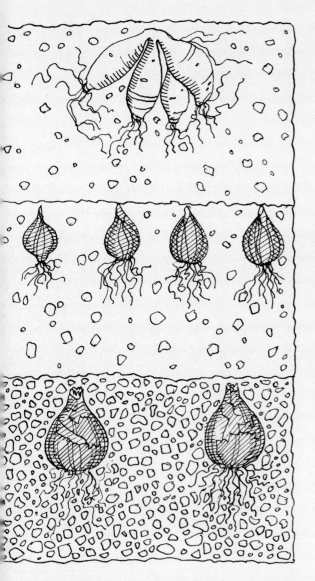

Creating a Dahlia Bulb Lasagne

This is the method we use for a multi-layered bed of bulbs and tubers with narcissi, Dutch iris, hyacinths and alliums.

* Dig a trench that's 30cm (12in) wide and 30cm (12in) deep. The length will depend on how many bulbs you have ready to be planted
* We place 3-4cm (2in) of grit in the bottom to help drainage on our heavy clay soil and I'd advise doing this on all but the most free-draining soil.
* Plant the narcissi into the grit, allowing twice the width of the bulb as spacing. This allows room for naturalising without a need for urgent dividing.
* Create a mix of 1 part grit to 3 parts soil and backfill to 4cm (2in) over the narcissi.
* Add a layer of Dutch iris, hyacinths and alliums at this level, around the bed edge.
* Backfill with the grit and soil mix (as above) to 4cm (2in) over the iris bulbs.
* Add the dahlias at 75cm (30in) spacings, just below the soil level.
* If we have dahlias that we're lifting in autumn, we mark each dahlia with a huge label so we know exactly where they are. Or if planting dahlias in spring, the label reminds us where we want the tubers to go in.
* Mulch over the dahlia crowns to at least the depth of 10cm (4in) in a mound.

Planting Bulbs in Pots

As containers become empty of summer bedding plants, it leaves us with lots of pots for planting up with miniature bulbs, which will bring great cheer come February and March, before the garden really gets going.

Plant some mini iris (*I. reticulata* and *I. histrioides*) and they will be in flower in January (if undercover) or February (if outside). And a few weeks later, the velvet green and black snake's head (or widow) iris (*Iris tuberosa*) will flower if planted this month. This is one of my favourite bulbs, sultry, exotic and sweetly scented. See page 30 for my selection of favourite winter-flowering irises.

Then there's crocus – I love *C. tommasinianus*, *C. vernus* and *C. chrysanthus* hybrids. At this time of year, I also plant the smaller narcissi (such as 'Xit', 'Minnow' and 'Canaliculatus') and species tulips (such as 'Little Beauty' and the spidery and elegant *T. acuminata*).

❀ Fill pots ¾ full with non-peat-based compost.

❀ Cram the bulbs into the pot so that they are close together, but not touching. To get something looking spectacular, I plant at twice the density I would in the garden, so rather than twice the width of the bulb, space them out at just one width of the bulb. And if I have plenty, I plant in a double layer.

❀ They all need a spell in the cold (at a temperature below 10°C/50°F) for 12–15 weeks. Put them in a cold frame, or if you don't have one, just leave them outside. They need this cold phase to develop their root systems before the demands of flowering.

❀ Check them every couple of weeks to see if they're starting to sprout. When you see them sprouting, you can move them somewhere warmer (such as a greenhouse) if you want to force them on for earlier flowering. Otherwise, simply leave them outside and they will flower as the weather warms.

November
& December

The gratitude I have for plants that are still going strong at this time of the year – and those just getting going – is enormous. They are beacons of colour and cheer. Whether growing happily outside or picked and brought inside, I want to pat them on the back to say, well done and thank you.

In November, and until we have our first hard frost, we are still picking a few dahlias. We have to hunt around to get good ones, but can still find them. And our chrysanthemums have come inside the greenhouse, so we can pick armfuls of those until Christmas.

Equal to these two plant heroes at this moment of the year is cobaea. We often have arches flowering into December, which are abuzz with bees whenever there's a hint of sun. I pick a vase a week, cutting the new shoots with tendrils, as well as the cup-and-saucer flowers; it's such a good combination of shapes. I've mentioned *Cobaea scandens* already in August (with the other climbers), but with so many plants now fading, it's worth a cheer for making its presence felt right to the end of the year.

The shoo-fly plant (*Nicandra physalodes*) is another stayer. Gardeners can be a bit disparaging about it. Usually brought into the garden via bird droppings, it appears in random places (very common in communal gardens), and as a self-seeder it tends to be dismissed. I feel differently. Blue tits and goldfinches love the seeds, burrowing tiny holes in the seedpods to get at them. One plant can create a metre-wide balloon of lantern seedpods in blue, green and black, and they don't lose their colour until Christmas. They look good early in the autumn with dahlias, and by the end of the year they complement the orange Chinese lanterns (*Physalis alkekengi*), crab apples (*Malus* 'Dartmouth') and hawthorns (*Crataegus persimilis* 'Prunifolia'). These are all invaluable for indoor decorations and wreaths.

Then there are the rudbeckias, reliable and able to put up with the cold and wet. The cut-and-come-again annuals, *R. hirta* 'Cherry Brandy' and the 'Sahara' mix, are, amazingly, still flowering. We often combine them with a June sowing of antirrhinums. Sown as late as this, the antirrhinums are still full-tilt in November.

Pelargoniums, particularly the scented varieties and the ones with bold, strong leaves, are also valuable at this time of year. My all-time favourite is 'Attar of Roses', its essence is used to make delicious rose geranium oil in the perfume industry. All pelargoniums are wonderful, whether for picking or for bringing in as houseplants.

We are passionate about sustaining a large wild bird population here, so it's important to start to feed our garden birds as the rosehips, haws and seedheads become thin. We make nice-looking, simple bird feeders out of dogwood or willow cut from the garden. One of our gardeners is a dab hand at this and she weaves them in

Below A mid-November vase of *Rudbeckia hirta* 'Sahara' and the bottle-brush foliage of *Chenopodium botrys*.
Opposite A vase of *Cobaea scandens* with vines picked as well as flowers to give the arrangement shape and scale.

double hoops, tied together in a sphere. We then hang a lard and seed block in the middle. We have these hanging in many of our smaller trees in the garden, as well as at our windows.

On a good day, each one will have seven or eight tits (blue tits, coal tits, great tits), as well as a couple of nuthatches on it, at any one time. Below there may be robins, a thrush and a blackbird or two, picking up the remnants the other birds drop to the ground. They give the garden life, they give us pleasure, and I'm certain this is why aphid infestations and slug and snail plant annihilations are becoming a thing of the past here.

We also make a few bird-food wreaths, with lardy fir cones and plenty of moss. And every year we put up more nest boxes, hang out wool fleece for bird nesting material, make bug homes, or just leave log piles for wildlife to hide in. I never used to take this seriously, but having seen the enormous wildlife benefits, I'm now a huge devotee.

As Christmas approaches, I am also on the lookout for interesting seedheads and flowers that I can dry. There are two grasses, the greater quaking grass (*Briza maxima*) and North America wild oats (*Chasmanthium latifolium*), which are both easy to grow, beautiful and delicate when growing or dried – they are the opposite of bling, even when sprayed gold. The chasmanthium is still going, but we harvest the briza earlier in the year and dry it. I love to fill small bottles with a handful of each to hover over our Christmas table.

Hydrangeas are also useful, the full and fluffy 'Madame Emile Mouillère', pyramidal 'Limelight' and smaller, 'Little Lime', as well as the skeletal, delicate, climbing hydrangea. Heads of all of these are brought in now to be dried.

From November through to February is the time for forced bulbs. We grow a big range – amaryllis, hyacinth and paperwhite narcissus – and we got going with some of these in September (see page 298).

As Christmas approaches, we also spray our alliums, harvested back in July and August, they've been hanging to dry in the Perch Hill classroom since then. Left out in a barn where the air is damp, we find all our pods and seedheads tend to go mushy and blacken with mould – they really benefit from a spell in the dry.

Christmas doesn't need to be all conifers, plastic and fakery. It can also be about plants growing and flowering, or foraged from the garden or hedgerow, with the outside brought in.

Pelargoniums

Still flourishing at this time of year are many of the pelargoniums. We've grown pelargoniums here for years, mainly in pots to arrange in groups or lines around the garden, but recently we started growing them as border perennials. It's warmer here than in other parts of the country, so in our sheltered areas the tougher pelargoniums survive the winter.

I have always loved picking them for their foliage and use them in hand-tied bunches and vases through the summer and autumn. I also add the foliage into the moss bases (or water vials within the moss) for our Christmas wreath. They last three weeks in water and lots of them have great perfume – all you need do is squeeze a leaf and delicious scent is released.

We use them in the kitchen, too, as they also provide unusual, delicate flavour. We make tisanes with their leaves and lots of cordial, and I use some for flavouring apple pies and crumbles. The great garden chef, Elizabeth David, was an avid user of the pelargonium leaf and that's recommendation enough.

A few years ago, with our ever-increasing demand for pelargoniums (particularly the scented ones), I decided to line the grass path down the bank with a mix of three exceptional varieties for cutting and cooking: I chose 'Pink Capitatum', 'Sweet Mimosa' and the great 'Attar of Roses'. We had taken enough cuttings the previous autumn (see page 266), to create a mini hedge on both sides of the path.

We planted them out in May, and over the next five months, we harvested buckets, cutting all the more ruthlessly as we felt winter approaching – if the frosts were going to have them, we might as well hack away. And the frosts did indeed scorch the top 15cm (6in) of the hedge one morning in the middle of December. Normally, we would have dug the whole lot up and put them on the compost heap, and we would have started again with cuttings in the polytunnel the following spring. But at that time, we were busy with other things, and so left the fifty or so plants where they were.

Scented-leaf pelargoniums lining the path of the veg slope, including 'Attar of Roses', 'Pink Capricorn' and *P. quercifolium*, all invaluable for early-winter picking.

The following April, after a mild and sunny week, I noticed they were budding again. Beneath the brown tops, the stems were reshooting. It was almost as if the frost had pinched them out and they were branching and pushing the dead foliage away as the new growth erupted from below. What a triumph! We cut them back a bit. They went on growing, bigger, better.

That's how we grow many of our pelargoniums now, adding more every year and thinking of them as part of our border – as well as pot plant, cut flower and foliage – repertoire.

On the cut-flower front, we've added 'Lady Plymouth', the upright, *P. crispum* 'Variegatum' with variegated leaves and *P. crispum* 'Cy's Sunburst', with powerful citrus fragrance. They all look lovely, smell delicious and last brilliantly in water.

Pelargonium 'Aurora' almost doesn't stop flowering. I bring it inside for winter, out of the frost, and on and on it goes.

We still have the luxurious crimson velvet 'Lord Bute', as well as varieties like the drought-resistant, ever-flowering 'Aurora' and 'Shrubland Rose', planted on their own in pots as living table centres. We use some pelargoniums for their handsome, scented foliage in pot combinations too: 'Chocolate Peppermint' is my current favourite in this role, bright green and felted with a soft crimson mark towards the centre of each leaf, and the species *P. quercifolium* is a beauty.

I'm a massive fan of the downy, scented, silver-leaved, *P. tomentosum*, which smells strongly of mint. It's exceptional because it grows equally happily in sun or shade. We have a whole water trough planted up with this against the north face of our barn every summer. We also grow it in a couple of large pots that can then rotate from the garden to the house, as they become houseplants for winter.

It makes sense that *P. tomentosum* in particular makes a good indoor plant, as the conditions in a house are much the same as in dry shade. I love it sitting indoors all winter in a terracotta pot on a glazed water tray positioned on our deep window ledges. There's one thing to note about having them inside: water them really sparingly. Over, rather than under-watering, is usually what kills them.

Both the ivy-leaved pelargoniums, 'Surcouf' (in brilliant pink) and 'April Hamilton' (in crimson), also make excellent houseplants, trailing down the side of the pot from a high shelf or window ledge and remaining in flower through winter.

I decorated the whole Christmas table with 'April Hamilton' a few years ago. I scattered patty pan squash 'Custard White' (they're like flying saucers) and Christmas baubles down the middle of the table, adding pots of 'April Hamilton' between them, as well as crimson pumpkin-shaped candles all through. This arrangement was virtually zero maintenance, mostly home-grown and lasted for weeks from the start of Christmas until well after New Year, with only an occasional watering of 'April Hamilton'.

For many reasons, it's hats off to pelargoniums from November until spring.

Best of the pelargoniums

Our collection of pelargoniums is ever-expanding, as are the ways in which we use them. We plant them in pots, large and small, and use them as late-season fillers in our borders – and they migrate inside as wonderful houseplants for winter. These are my favourites.

1 *Pelargonium* 'April Hamilton' (Ivy-leaf)
A crimson, ever-flowering, shiny-leaved variety that tumbles. I love this in alternating pots with its bright pink relation, 'Surcouf'.

2 *P.* 'Attar of Roses' (Scented-leaf)
No fragrance-obsessed gardener's collection is complete without this one. With pretty flowers and bright-green leaves that are strongly rose scented, it is the Scented-leaf pelargonium classic. It's a quick and easy grower, almost cut-and-come-again.

We use its flowers in arrangements and the leaves for tisanes and cordial. We pick the leaves regularly, boil them briefly in water and then steep them overnight to make into a syrup. We also add citric acid (to preserve it) and fresh lemon juice to make one of the best ever cordials.

3 *P.* 'Aurora' (Zonal)
If you like full-on colour, go for 'Aurora', which has pompom heads in the brightest fuchsia. It never stops flowering. Don't think it vulgar, think it glorious.

4 *P.* 'Chocolate Peppermint' (Scented-leaf)
This has indistinct pink-mauve umbels of flowers, but very dramatic felted-leaves and a distinctly minty fragrance. This makes it very handsome as foliage for a pot and for a vase.

5 *P.* 'Clorinda' (Scented-leaf)
The tallest-growing pelargonium we grow, with large, clear, brilliant pink flowers from very early in the year until late. If not cut back, this flowers from February in our greenhouse, with new cuttings flowering from June.

6 *P.* 'Lara Starshine' (Scented-leaf)
Brilliant pink, and hugely long-flowering from spring to winter, this is garden designer and friend Pip Morrison's favourite – and he has a great eye for a great plant.

7 *P.* 'Lord Bute' (Regal)
One of the richest and darkest of the very showy Regal types, and it flowers longer than most in this group. It gets better the bigger it grows, so I store it in its pot from one year to the next, out of the frost.

8 *P.* 'Orsett' (Scented-leaf)
Flowering from April or May to the end of the year, this is one of our cheery garden and pot stalwarts. I love it for its reliability and durability – we can easily keep plants from one year to the next.

9 *P.* 'Shrubland Rose' (Scented-leaf)
A brilliant pink and unbelievably long and prolific flowerer on truly minimal TLC. We put this in our furthest-flung pots by the Chelsea shed in the Annual Cutting Garden. We know the pots

there have a tendency to get a little overlooked compared to those in the heart of the garden. Even there, 'Shrubland Rose' does fine. We had a great combination of this with the equally tough-as-old-boots (and just as showy), *Cuphea caeciliae* and *Argyranthemum* 'Granddaisy Deep Red'. That's one of my favourite pot combinations.

10 *P. quercifolium* (Species)
This species has a strong cedar scent and a crimson splotch at the centre of every leaf and its Latin name means 'oak leaf'. It's excellent for picking and lasts nearly a month in water. Good for pot foliage, too.

11 *P. sidoides* (Species)
Totally hardy with us, this has survived outside now for ten years. I love its crimson, delicate-looking flowers in contrast to its rouched skirt and silver leaves.

12 *P. tomentosum* (Species)
Strongly scented with minty-geranium perfume and the softest texture of any plant I know – this is the stuff beds in heaven are made of. It's a rare pelargonium as it thrives in the garden in dry shade and makes a great houseplant. It's shy to flower and then the flowers are small and delicate.

A November dawn on the veg slope with *Verbena rigida* still looking good, and plenty of Florence fennel, kale 'Curly Scarlet', chard and chervil.

Amaryllis

In our house, *Hippeastrum* (commonly known as amaryllis) is as much a sign of Christmas as the best bauble. Without at least one good pot of three whopper amaryllis bulbs, this time of year would not be the same. From December right through until Easter, they are one of our guaranteed indoor deliverers of colour – and they don't need much TLC.

For many reasons, amaryllis have stolen the show over poinsettias as the Brits' favourite Christmas plant, and that's good news. The growing market means there are new and exciting varieties being bred all the time. It can be frustrating, as when new ones are launched, there is a tendency for old ones to be dropped. But as with sweet peas, I now get less hung up on individual cultivars and concentrate instead on the colours and forms that stand out.

Because some amaryllis are super-chunky, it's easy to think they're all coarse, but that couldn't be further from the truth. Look at 'Emerald', as fine and delicate as you could hope for, purposely made to look as though you've just come across a clump in its ancestor's jungle habitat of Peru.

When I visit Keukenhof in Holland looking for new tulips every April, I always look at the amaryllis breeders' stands – that's where I spotted the most incredible luscious crimson, 'Mandela', with smaller flowers than usual and definitely the darkest I've ever seen. Its richness and scale gives it class, and since trialling it here last winter, I now know its emergent leaves are glamorous too, washed with crimson.

As far as colour goes, I tend to avoid any bright, pillar-box reds (such as 'Red Lion'), which are commonly forced for Christmas – and I have reservations about the new full-on salmon-coloured hybrids. Both of these look good when newly emerged from bud, when the outer petals are still washed with green, but once they're fully out, I find them too garish.

At the other extreme, subtle and delicate, I love the soft pale greens. There are varieties such as 'Lemon Lime' and 'Lemon Star',

Amaryllis 'Misty', which is best en masse, rather than as a single bulb.

Amaryllis 'Green Magic' with moss
and lichen on the blackthorn twigs
that are supporting its stems.

with a white base, but my current favourite is 'Tosca'. Here, the
base colour has been dusted with crimson pigment, denser on the
top three petals than the lower. Those are the colours we grow here,
plus some pure whites.

We have, through the years, grown different flower shapes.
You can't help but be impressed by the large, blousy, tuba-shaped
varieties, as big as any flower you'll find any time of year, let alone
the winter. I know these aren't to everyone's (including our head
gardener's) taste, but I love them and would miss having the
brouhaha they guarantee.

There are also more delicate and interesting forms, such as
'Chico', 'Tango' and 'Emerald'. The flower-life of the large amaryllis
does seem to be longer, but even with the small ones, each
individual flower lasts well over a week. There will be several on a
stem (which open in succession), and several stems from one bulb,
so they put on a good, prolonged show.

Rather than planting just one amaryllis bulb, if possible, go for
a group of three. It's expensive the first time, but I have kept our
bulbs for years and now have quite a number for filling our tables,
shelves and plant stands to tide us over through these colour-lean
months. They need replanting every September (for Christmas)
and October (for early in the year). See page 300 for more on this.

I want to make them look more than a weird-stick-erupting-
from-its-pot, so rather than leaving them bare, I love making
mini landscapes with moss and lichen branches from the woods
and willow from the garden, which I weave into nest supports.
Amaryllis needs its heavy flowers staked, so this is functional as
well as beautiful.

We grow the elegant-stemmed willow, *Salix purpurea* 'Nancy
Saunders' at Perch Hill, partly for this staking reason. It drops its
leaves in October and is then bare and pliable, easy to weave into
supporting hoops. The flowers usually continue well into the New
Year, when the willow, even cut, starts to form its soft silver pussy
willows, giving our amaryllis pots a fresh dimension. I then scatter
dried leaves with crisp, clean shapes (such as sweet chestnut, oak
or maple) over the moss.

With the amaryllis leaves and flowers growing week by week,
we have a whole new scene to enjoy, which in December, we crown
with small-scale Christmas baubles threaded on to the willow.
It's a key part of our 'growing' Christmas.

Best of the amaryllis

People tend to either love or hate amaryllis, but increasingly there really is an amaryllis for every taste. There are elegant, narrow-petalled Spider types and ones with vast, trombone flowers. These are the ones that have all done well for us.

1 *Hippeastrum* 'Alfresco' (Double)
This is a double white, shorter than most, with stems only 45cm (18in) tall. I think this is in danger of looking top-heavy, but it's good if you have limited space.

2 *H.* 'Chico' (Spider)
Tropical-looking and exotic, I love this one. It has green and deep-red spidery flowers and looks more like an orchid than an amaryllis. Its delicate flowers go over more quickly than the rest, so it's worth keeping this one cool and lightly shaded while in flower.

3 *H.* 'Emerald' (Trumpet)
Lovely, subtle and delicate, with white-green flowers that have the faintest crimson veins and pencil-thin edge.

4 *H.* 'Giraffe' (Trumpet)
Smaller flowers than the classics, with pretty green and red stripes. Beautiful, but flowers are not very long-lived.

5 *H.* 'Green Magic' (Trumpet)
Outstanding, with soft green petals dusted with crimson; like 'Tosca' but with less claret.

6 *H.* 'Hercules' (Trumpet)
Its flowers are huge and an incredible pink-red, with five or six flowers to a stem. I didn't immediately love this one, but it grew on me.

7 *H.* 'Lemon Lime' (Trumpet)
A lovely green, the colour spreading out over the white petals with green veining right to the petal edge. 'Lemon Star' is similar and equally worth a mention.

8 *H.* 'Mandela' (Trumpet)
The darkest of any amaryllis I've ever seen, and petite with it, this newly bred variety is now my favourite in the rich crimson palette. It has wine-red staining over the stems and nascent buds.

9 *H.* 'Mont Blanc' (Trumpet)
The most spectacular white we've grown: huge ivory-white flowers with a green throat. Also, lots of flowers on one stem and a good succession of stems, so it flowers for months.

10 *H. papilio* (Trumpet)
True green with the central section to each petal striped with crimson. This, as its name implies, looks like a huge tropical butterfly.

11 *H.* 'Pink Rascal' (Sonatini)
Shaped like a day lily, with stems just over 30cm (12in) tall, this delicate pink is happy flowering outside twelve weeks after planting, not in winter, but in spring. 'Red Rascal' is the scarlet version, like stained glass with an elegant green-crimson stem. These Sonatini are from a cross between the indoor *Hippeastrum* and the pink-trumpeted garden plant, *Amaryllis belladonna*.

12 *H.* 'Royal Velvet' (Trumpet)
This is the brightest of the reds, but still lovely, with a crimson wash over the scarlet.

13 *H.* 'Tosca' (Trumpet)
This, together with the similar, softer-coloured 'Green Magic' and *H. papilio*, are my three favourites. I particularly like the green and crimson mix you get with all three.

Practical
November & December

With winter weather, it's quite a relief that many of the jobs for this time of year can be done inside, though not all. The main outdoor job for November and December is planting out all of the tulips. That's no small matter here: with our pots, borders and trials, we usually have nearly 20,000 tulip bulbs to go in.

We leave our dahlias in the ground, so they need to be cut back and then protected with a good dollop of mulch over each crown. That's time-consuming, but so much easier than having to dig the whole lot up, only to replant them again in May.

Inside, I make a tiered showstopper using paperwhite narcissus, and I make a wreath or two for the doors, as well as prepare the dried alliums and seedheads that were harvested in the middle of summer, spraying them to decorate our tree and table for Christmas.

Planting Tulips in Bulb Lasagnes

For dense and flowery pots of tulips, I plant them in what the Dutch call a 'bulb lasagne', layering them up one on top of the other. The emergent shoots of the lower level just bend around anything they hit sitting over their heads and keep on growing.

In our large long toms and Danish terracotta pots, I plant fifteen bulbs in each layer, so there are forty-five bulbs in a triple decker. Aim to choose a mixture of tulips that have stem lengths that will work well together and flowers that come all at the same time.

I often add a final fourth layer of iris or crocus, just poking them in beneath the compost surface. *Iris reticulata* is a wonderful way to start the whole procession off in February, just when we need a little cheer, and the elongated foliage that comes later makes a good foil to the tulip's flowers (see page 30).

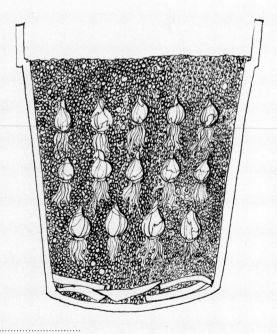

Planting tulips in a bulb lasagne. The black inner pot slots into a dolly tub (see p100) to make colour succession easier to achieve. We can then whip it out and replace it with late-spring flowerers once the tulips are over.

❧ If you're planting a mix of bulbs, the largest and latest flowering bulb goes in deepest, moving to the smallest and earliest in the top layer. The larger the bulb, the more starch it has for the longer journey into the light, when the leaves start to photosynthesize and make their own food.

❧ If you're planting only tulips to flower at the same time, just mix them up together in a bucket and plant randomly.

❧ Plant the bulbs slightly further apart than you would if you were planting them in a single layer. Around 2-3cm (1in) is the right sort of spacing.

❧ Plant the bottom layer at a depth of 30cm (1ft) and the middle at 20cm (8in). The top layer of bulbs should be planted 10cm (4in) below the compost's surface.

❧ Once all the layers are planted, finish with a sprinkling of chilli powder or chilli flakes generously scattered. We have found this pretty successful at keeping grey squirrels, rats and mice at bay.

Planting Tulips in Borders

Planting deep is good with tulips. It seems to make the bulb last longer and flower for longer. At a lower level in the soil, the bulb does not get warm, and it's warmth that encourages bulbil formation and reproduction. Too much of this and the mother bulb gives her energy up to the bulbils, which are not yet of flowering size. The mother bulb can then no longer flower and the young ones are still too small, so the bulbs are blind (and you get no flowers). Planting deeply avoids this. It also helps protect against tulip fire infection. Also, buried deep, it's harder for squirrels and other rodents to dig them up.

For a non-back-breaking way of planting bulbs deeply in borders and grass, we use a stainless-steel bulb planter – not one with those hopeless short handles that give you wrist pain for weeks, but one with a proper shaft on it like a spade.

* A bulb planter has a round corer at the bottom. Push the corer into the ground with your foot and remove a cylinder of soil or turf.
* On heavy soil (like ours), drop in a handful of grit or spent compost.
* Plop in the bulb.
* Move onto the next. As you push into the ground for the second hole, the first core pushes out of the cylinder in the bulb planter. Use this to replace the soil over your first planted bulb. Firm down.
* Drop in a handful of grit or spent compost into the second hole. Then plop in the bulb... and so on.

Planting Tulips for Cutting

When I want to plant large quantities of bulbs for picking, I find there are too many to plant each one in its own individual hole, so I dig trenches and pack the tulips in them in a double or triple layer.

* Dig out a trench 30cm (12in) deep.
* Scatter 3-4 cm (1-2in) of grit over the base.
* Plant the tulips at double the width of the bulb. I used to really pack the tulips in the trenches, but tulip fire has put a stop to that. Once the bulbs are growing at full tilt, they need good air circulation between the flowers and leaves. If they are jam-packed, it creates the perfect humid conditions for any fungus to thrive. Good spacing is key.
* Create a mix of 2 parts soil to 1 part grit and backfill with this to the depth of about 15cm (6in).
* Add another layer of bulbs and cover with the soil/grit mix (as above).
* We often add companion plants over the top – wallflowers and honesty are both great, as well as polyanthus, pansies and forget-me-nots.
* Protect against rodent attack with chilli flakes scattered densely over the soil surface.

Winter Dahlia Care

Lifting dahlias

If you live in a frost pocket you will need to lift your dahlias from your borders. Similarly, if you have dahlias that have been growing in pots, the tubers need to be brought in for the winter.

❧ Cut the plants back to 10cm (4in).
❧ Dig them up or knock them out of the pot.
❧ Dry off the tubers indoors for 1-2 weeks. Do this by standing them upside down to drain the stems. We put ours in crates in the polytunnel.
❧ Remove any remaining loose soil and then pack the tubers into boxes of dry compost.
❧ Store so that they are frost-free in a shed or under the bench in the greenhouse. You can keep them in a cellar or cold room indoors, but avoid warmth or they may dry out.

Mulching dahlias

If (like us) you are able to leave the dahlias in the ground, mulching is key to their survival.

❧ After the first hard, leaf-blackening frost in November or December, cut the dahlias back to within 10cm (4in) of the ground.
❧ Mulch deeply, tipping a bucket of compost or green waste over the head of each one. Dome it over the crown. This will insulate the tender tubers through the winter and direct the rain away so there's less danger of rot.

❧ Label clearly so you remember what's what.
❧ You can push the mulch aside once the weather warms up by late April or early May.

Saving and Storing Amaryllis

The older and bigger the bulb, the more flowering stems you'll get, so it's worth going to the trouble of nurturing these mini football bulbs. They're expensive, too, so this is what we do.

❧ Once the amaryllis are flowering, continue to water and keep them out of direct sunlight. Once in flower, try to keep them slightly cooler (10-15°C/50-60°F).
❧ As each flower fades, cut it off individually, leaving the others to bloom.
❧ When every flower on one stalk is over, cut it off to just above the bulb nose. With a large bulb, there should be at least one or two more flowering stems to come.
❧ After flowering is completely over, continue to feed the plant (once every six weeks) and continue to water it, until the leaves begin to turn yellow.
❧ At this stage, cut the leaves back to about 6cm (2½in).
❧ Don't repot immediately. First, keep the bulb somewhere dark and cool (10-15°C/50-60°F), allowing it a dormant period of ten weeks.
❧ After the dormant period, repot in fresh compost, trying to minimise root

We lift our dahlia tubers from pots and dry them upside down for a couple of weeks before storing in hessian sacks.

disturbance. Start watering again 8-10 weeks before you want them to bloom.

❧ Bulbs older than two years will produce offset bulblets. These may be left attached to the mother, creating an amazing show, but it's best to remove them carefully to give yourself more plants. Knock the whole bulb out of its pot and replant all offsets in their own individual pots. These little bulbs will take two years before producing their first flower, but it will be a proud moment when they do.

❧ Amaryllis is a tender bulb, so needs to be grown inside and remain frost-free in the winter. But once the frosts are over and the nights no longer cold, they can be moved outside until the end of summer.

❧ It's worth knowing amaryllis flower naturally from March to May and are ideally planted at the end of a dormant season in November or December. If planted in September (see page 300) they flower for Christmas, but you can't repeat this early planting year after year, as the bulb may get starved and not flower as well.

A three-layered 'cake' of paperwhite narcissi supported on hazel branches nearly a metre high.

Paperwhite Narcissus Table Centre

Years ago, a distant cousin sent me a picture of a sort of tiered cake, made not from sponge and sugar, but scented narcissi bulbs. Creating them had become a tradition in the family, passed down from one generation to the next, and I've been creating something similar in the winter ever since. My cousin used the orange and gold *Narcissus* 'Grand Soleil d'Or' in small bowls, but never one to do things small-scale, I tripled its size and used a huge Spanish bowl as the base,

and the lovely, paperwhite narcissus. I can't recommend them more highly.

Paperwhites are the easiest and quickest bulbs to flower inside, but to have them in flower by Christmas, you need to plant them in early November, giving them at least seven to eight weeks. These days, I make several of these table centres, planting them a couple of weeks apart to give a good succession of colour and scent through winter. Admittedly, to start with, there is only a nest of narcissi leaves and silver birch twigs, but even this, in a beautiful bowl, holds its own without flowers.

If it's too early to start thinking about Christmas in early November, or I'm too busy, I sometimes cheat and buy narcissi that are already potted and growing, transferring them into my own pots a week or two before Christmas.

❦ You need two pots of different sizes. The first should be as large a pot or bowl as you can fit in the middle of your table to form the base of the arrangement, with a smaller one for stacking on top. You can add a third if you have space. Narcissi bulbs are large, with an extensive root structure, so deep pots are ideal.

❦ You can plant the bulbs initially in plastic pots and then move them into your final table centre as they come into flower, or do them straight into their final planting pot.

❦ Plant the bulbs about 2-3cm (1in) apart into bulb fibre. You can also use loam-based compost, lightened with some grit. Plant the bulbs just below the soil surface.

❦ Narcissi need a spell in the cold to flower well, with a temperature below 10°C (50°F). Most varieties take sixteen to eighteen weeks from planting to flowering, but not paperwhites. These are super-speedy and need only seven to eight. And they don't require a period in the dark, either, which makes them doubly easy-going.

❦ Keep the compost moist but not dripping wet.

❦ Once they really start to shoot, with leaves up to 20-25cm (8-10in), bring them into the warm.

❦ If they're still in plastic pots, transfer them into your final pots. Pack the flowering bulbs in as thickly as you can.

❦ Poke a handful of silver birch or hazel twigs into every pot layer to support them, spacing them at about 10cm (4in) around the outside of each pot, making sure you've pushed them right to the bottom.

❦ Bend the twigs at right angles, 15-20cm (6-8in) from the compost surface; they should be pliable enough not to break. Then start twisting and weaving each onto its neighbour, doing so horizontally.

❦ When you arrive back at where you started, bind back over to secure the end of the last bit of twig. This looks lovely and staves off collapse, keeping the whole thing looking good for longer.

❦ As a final touch, drape the twigs with silver and clear glass baubles and surround the whole thing with a halo of candles on the table.

❦ Once it's all over, bear in mind that the paperwhites are not hardy (go for N. 'Avalanche' and 'Geranium' if you're looking for a hardy variety), so once they've finished flowering, leave them in their pots for next year, or dry them off and repot them again next autumn.

Making a Christmas Wreath

If the autumn and winter gales are yet to flatten our garden, I get out there quick with my florist's scissors and harvest the last of any seedheads and pods before they're swept away. Hydrangeas are usually just about still going and the last Japanese anemone seedheads, and the grasses, including miscanthus and panicum. I also love collecting bracken, clematis seedheads and the first pussy willow and hazel catkins.

Spray any or all with supermarket hairspray as this stops them being so brittle and has the same effect (at half the price) as an art shop-bought fixative. Alternatively, go for a coat of spray paint.

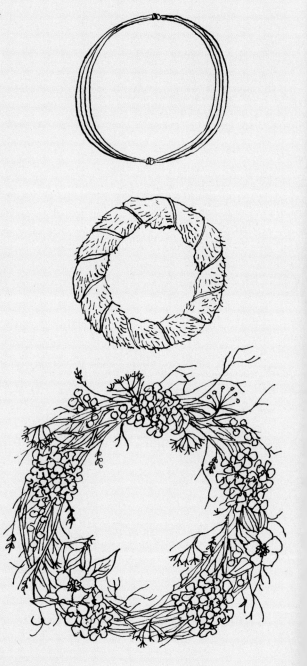

- ❧ A couple of weeks before wreath-making time, I pick any good-looking hydrangeas and, if it's been dry (unlikely!), I float them overnight in a bath of cool water.
- ❧ Then I put the whole bunch in a bucket for a couple of weeks with 2-5cm (1-2in) of water in the bottom. As the stems suck the water, the flowerheads dry and keep their colour. If you want them to dry retaining their texture, but not their colour, add a drop of glycerine into their water.
- ❧ I also pick wild clematis, old man's beard (*Clematis vitalba*), and the fluffy-headed varieties of garden clematis, such as *C.* 'Bill Mackenzie'. With these, I use the glycerine technique, so that the seedheads remain intact and don't shed all their fluff.

A dried hydrangea, eryngium and *Smyrnium perfoliatum* wreath hanging on the inside of the barn door.

Place the stem ends in a jug with just a little water in the bottom and a drop of glycerine. Leave for a couple of weeks until dry.

❈ To create a whopper wreath, I use fencing wire in a triple circle, bound top and bottom, rather than a ready-made frame.

❈ If I want to add flowers or fresh foliage (such as chrysanthemums from the greenhouse, or some sprigs of rosemary for a rich colour and scent), the first step is to pad the frame. I use moss scarified from the lawn. Lay an even and generous layer over the frame, the same amount all round. Fix this on with florist's wire (thin wire on a spool), binding it quite tightly.

❈ Increasingly, I like my wreaths with dried and everlasting things that don't need moss, so I often skip this stage.

❈ Add a loop of wire with which to hang up the finished wreath and mark this by tying on a piece of ribbon.

❈ Whether you've got a moss base, or just wire, you can now cover the whole thing with silver birch or dogwood twigs.

❈ Now start adding whatever you've found and harvested. We pick fennel heads, clematis, bracken and grasses, a mix of which covers the wreath base evenly. We also use climbing hydrangea vines, which give great shape and their heads are skeletal and elegant.

❈ Then add a few freshly cut hellebores. The stem ends can be placed in small water vials hidden amongst the wreath.

Hang your wreath from the door (you may need to knock in a tack or small nail). If you've used a moss base, rehydrate it in a sink of cold water overnight every week or so to keep everything going longer. I usually add battery wreath lights to light the whole thing up when it's dark. It acts like a beacon of welcome when the nights are so long.

Index

Figures in **bold** refer to main entries

Acknowledgements

This book has been like an octopus involving so many different people over so many years, but I particularly want to thank a special few.

For their general kindness and encouragement, thanks to Rosie and Molly Nicolson, Liam Ashmore, Tania and John Keeling, Anna Canetty-Clarke, Caroline Nevile, Arthur Parkinson and my agent Caroline Michel.

At Bloomsbury, thanks go to Richard Atkinson and Natalie Bellos, who guided me through making books over many years, and more recently to Rowan Yapp and Kitty Stogdon.

For continuous garden and plant inspiration, thanks to Pip Morrison, Gary Newell, Josie Lewis, Arthur Parkinson, Faith Raven, Carien Van Boxtel and Dicky Schipper.

For encouraging Jonathan and me to do the practical shots so important in this book, thanks to Ceri Thomas from Which? Gardening.

For incredible hard work and cheerful help in making the garden look so good every month of the year, thank you to Josie Lewis, Mary Pocock, Colin Pilbeam, Anita Oakes, Whitney Hedges, Jo Skinner, Henry Macaulay, Richard Lambden, Sophie Kusel and Adam Nicolson.

For encouraging readers along the way, thank you to again, Adam Nicolson, Matthew Rice, Lou Farman, Anita Oakes, Josie Lewis, Caroline Nevile and Kate Hubbard.

This book would not be a fraction of the object I feel truly proud of without the careful, concise and speedy editing and guidance of Zena Alkayat, the drawings of Helen Cann, the airy and clear design of Glenn Howard and the, to me, movingly beautiful photographs by my dear old friend and collaborator, Jonathan Buckley.

I've absolutely loved writing and pulling together this book – and that's largely down to all of you. Huge and heartfelt thanks.

Most of the plants included in this book are available from sarahraven.com

BLOOMSBURY PUBLISHING
Bloomsbury Publishing Plc
50 Bedford Square, London, WC1B 3DP, UK

BLOOMSBURY, BLOOMSBURY PUBLISHING and the Diana logo
are trademarks of Bloomsbury Publishing Plc

First published in Great Britain 2021

A catalogue record for this book is available from the British Library.

Library of Congress Cataloguing-in-Publication data has been applied for.

ISBN: HB: 978-1-5266-2611-0;
eBook: 978-1-5266-2610-3

10 9 8 7 6 5 4 3 2 1

Project Editor: Zena Alkayat
Designer: Glenn Howard
Photographer: Jonathan Buckley
Illustrator: Helen Cann
Botanical proofreader: Dr Ross Bayton
Proofreader: Helen Griffin

Printed and bound in China by C&C Offset Printing Ltd, China.

To find out more about our authors and books visit www.bloomsbury.com
and sign up for our newsletters.